CHARLES BONNET AND THE ORDER OF THE KNOWN

STUDIES IN THE HISTORY
OF MODERN SCIENCE

VOLUME 11

CHARLES BONNET

LORIN ANDERSON

Department of History, Portland State University

CHARLES BONNET AND THE ORDER OF THE KNOWN

D. REIDEL PUBLISHING COMPANY

DORDRECHT : HOLLAND / BOSTON : U.S.A.

LONDON : ENGLAND

Library of Congress Cataloging in Publication Data

CIP

Anderson, Lorin, 1946–
Charles Bonnet and the order of the known.

(Studies in the history of modern science ; v. 11)
Bibliography: p.
Includes index.
1. Bonnet, Charles, 1720–1793.
2. Naturalists—Switzerland—Biography.
3. Philosophers—Switzerland—Biography.
I. Title. II. Series.
QH31.B72A8 508 [B] 82-5320
ISBN 90-277-1389-8 AACR2

Published by D. Reidel Publishing Company,
P.O. Box 17, 3300 AA Dordrecht, Holland

Sold and distributed in the U.S.A. and Canada
by Kluwer Boston Inc.,
190 Old Derby Street, Hingham, MA 02043, U.S.A.

In all other countries, sold and distributed
by Kluwer Academic Publishers Group,
P.O. Box 322, 3300 AH Dordrecht, Holland

D. Reidel Publishing Company is a member of the Kluwer Group

Printed in The Netherlands

To Michael Reardon

TABLE OF CONTENTS

PREFACE

Charles Bonnet began his career as a naturalist, from an early age establishing a reputation as a careful observer. It is for those youthful observations, as well as for some suggestive speculations proposed relative to this field, that he is best remembered in English-speaking countries: regarding the taxonomic demands of natural history he refurbished the idea of a chain of beings; regarding the question of generation he marshaled evidence in support of preformation theory; and regarding the analysis of the physiology of the nervous system he advanced a theory that individual nerve fibers receive and retain specific sensations. Following his loss of eyesight in his mid-twenties Bonnet entered a more reflective period, turning to philosophy and pondering the nature of human understanding – considerations he had formerly disdained, but that now seemed a natural outgrowth of his reflections on nature. This essay focuses on the philosophical and psychological works of the later period, the period in which he wrote all his major books. By giving these writings a broader exposure it has been one of my hopes that Bonnet's audience would also be broadened, releasing him, so to speak, from the charge of historians of science so that he might find his way, in general books on the "Enlightenment", from scattered footnotes into the texts themselves.

With the exception of the first chapter, which sketches Bonnet's education and traces the development of his thought through a chronological summarization and analysis of his works, my approach has not been that of an intellectual biography, nor that of a synopsis of his writing as a whole. From several perspectives or vantage points, by tracing out the complexity and convergence of ideas constitutive of taxonomy, language, personality and the relational order in Bonnet's works, I have attempted to shed light not only on each of these subjects but also upon the network linking them together – and returning all of them, at different and unanticipated moments, to certain underlying assumptions and ambivalences. My intention has been to run augers into subjects whose own unities, demands, limits and qualms would offer insight into the web of Bonnet's thought. Not by surveying it, or by reducing it into a more manageable miniature: I definitely was not out to present an abridgment of Bonnet's system of thought – better to read Bonnet instead – but by probing what appeared to be the locus of significant ideas or concerns, and examining them in detail.

Each chapter, then, takes up an aspect of Bonnet's thought that is linked to all the others, but that at the same time stands by itself as a coherence and a disturbance. In the case of language, its place and functions are fixed, along with the needs those functions were expected to fulfill, all of which finally reveals the essential uneasiness upon which the entire edifice stands, since it provides a resolution to a problem that nonetheless persists, just out of visible range. And the same holds true for a system of denominating beings that makes certain undenoted adjustments when it can no longer comply comfortably with the continuum it advocates, with, for example, a classification of the animate and the inanimate in the same series, according to degrees of organization, while ignoring the determining principle of life; or for a concept of personality that hovers hesitantly between its various poles, which threaten ungovernability, dispersion and discontinuity on the one hand, and continuity at the cost of automatism on the other; or finally, for a representative order of thoughts never quite assured of its connection to the relational order of reality that it represents. The manner in which these areas overlap with one another in their assumptions, themes, metaphors and concerns gradually becomes clear, I think, and indicates the epistemological field upon which Bonnet's thought is ordered.

Throughout, the establishment of Bonnet's thought at this level has been, far more my direct concern than the determination of his place among various influences, traditions, debates and victories. I have had little interest in searching for Bonnet's precursors or in reducing his own role to that of a precursor, preferring instead to devote my attention to what he was actually doing, not behind or between the text, but specifically in the text. Why these words, why this consideration at this particular point rather than some other? Why this arrangement of chapters and topics? Why so much space given over to an apparently minor issue and then so little to other issues of "real importance"? I have tried to retain, to a certain extent, the language of Bonnet, the arrangement and weight he gave to particular issues, and his progressions from one topic to the next. Above all, I have tried to avoid simplifying Bonnet's thought where it was not simple, to stay with the problems that were problems for him, and to maintain the level of complexity with which he addressed them.

Certainly one of the great pleasures afforded me with the publication of this book is the opportunity to express in print my gratitude to my close friends of the Bonnet Institute, Mott Greene and Claude Singer, for their companionship and advice. For the interest he took in the manuscript and its publication, I would like to thank Paul Farber. For their various roles in shaping its form and arguments I should also like to acknowledge the efforts of Thomas Hankins and Rodney Kilcup. Finally, for its support, I wish to thank the Department of History at Portland State University.

ACKNOWLEDGEMENT

Excerpts from *The Order of Things: An Archaeology of the Human Sciences*, by Michel Foucault, translated by Alan Sheridan-Smith, are reprinted with the permission of Pantheon Books, a Division of Random House, New York, and Tavistock Publications, Ltd., London, © 1970. Chapter II of this book was originally published, in a shorter version, in the *Journal of the History of Ideas*, XXXVII: 1976, and is reproduced here with the permission of the editor.

CHAPTER I

INTELLECTUAL DEVELOPMENT

The life of Charles Bonnet was indefensibly dull; regularity characterizes its very essence. "As for my life", he acknowledges, "it has been nearly as uniform as the heavenly bodies".[1] Retiring and obedient as a child, unexceptional in his schoolwork, never brash during his years of maturation, unacquainted with wanderlust,[2] reclusive in youth and in adulthood, even he apologizes for the monotony of his uneventful existence.[3] With good reason, then, do his *Mémoires autobiographiques* aim at recounting "the history of my life, or rather, of my thoughts, for the very uniform life of an individual nearly always meditating or writing is scarcely anything except thoughts".[4]

The following biographical sketch will in general follow Bonnet's judgment in this regard, outlining the chronological development of his thoughts. His first interests were in microscopic observation, for which he received renown while still very young. He then turned from experiment and observation to philosophical and psychological speculation. My concern will be first with the elements involved in this shift of interest, and then with a consideration of some of the central issues raised in his major writings, since it is on these works that the following chapters focus.

Little time is spent by Bonnet recalling his childhood. He was born in Geneva on March 13, 1720, the son of Pierre Bonnet and Anne-Marie Lullin, and descended from a family that had been in Switzerland for six generations, having left France in 1572 following the Saint Bartholomew's Day Massacre. One family matter that would later affect Charles was his father's loss, in 1722, of a modest country residence in Thonex, the result of unpaid debts. This suspension made it impossible, according to Genevan law, for Pierre Bonnet to follow in the footsteps of his father and become a magistrate.

In elementary school, the young Bonnet won no awards; he remembers himself as not particularly attentive. Classical studies revolted him, and he found teaching methods vicious, their chief requirement being endless memorization of grammatical rules whose application remained elusive. These methods changed after he had passed through the system; nevertheless, his bad memories inspired him to return several times, as an adult, to considerations on methods of education. Boredom was not the only source of interference. Before he had completed his primary education, he had begun to suffer hearing loss and was withdrawn from school to continue his studies at home. This early seclusion was only the beginning of the pattern of retreat that marked Bonnet's life, which may be attributed both to his particular tastes and, increasingly, to the state of his health.[5] Under the guidance of a private tutor, Laget, Bonnet became more interested in Greek and Latin authors, and he began to read the *Spectator*, which he claims to have found exciting and instructive. His interest in studying increased, when, in 1735, at the age of fifteen, he entered the "*Auditoire de Belles-Lettres*". But it was the following year that he underwent an experience which he compares to the one undergone by Malebranche upon reading Descartes. By chance he picked up and read the first volume of the Abbé Pluche's *Spectacle de la nature*: "I did not read the book, I devoured it."[6] What caused him such excitement was the history of the ant lion, whose nature and industry so intrigued him that he wandered the countryside unti he had found one to observe for himself.

After finishing his studies in arts and letters, Bonnet entered the "*Auditoire de Philosophie*" in the Spring of 1736. And there he found two professors – Jean Louis Calandrini (1703–58) and Gabriel Cramer (1704–52) – both of whom he came to admire for their knowledge and teaching abilities. It did not take Bonnet long to decide that it was natural philosophy ("*la physique*") that he found most attractive, "doubtless through its relations to natural history, whose charms I had already begun to sense and which soon became master of my entire being".[7] The rational aspects of philosophy, on the other hand, were of no interest to him: "Only with great difficulty was I able to gain some understanding of abstract notions and arrange them in my brain. They were too fugitive or too ethereal for me; and when I believed I had something of a grasp on them, they escaped . . . . I was repelled by the throng of definitions, divisions and distinctions that rational philosophy presented, of which I discovered neither the merit nor the purpose."[8] So how was it that a young man, with so little interest in speculative thought, came to write an analytical essay on the mind? For Bonnet there is a pedagogical lesson here regarding the care that must be taken to search out talents and interests that lie dormant in youth, and that may never be revealed except by accident if unnoticed and untended by the teacher.

Among his own teachers, Bonnet took Cramer for his principal guide: "I consulted him as my oracle."[9] Cramer, who had been named mathematics professor in Geneva in 1724, would later edit the works of Bernoulli, an accomplishment that gained him d'Alembert's praise.[10] Recognizing Bonnet's preference for natural philosophy, Cramer had him read works in natural history by Swammerdam and Malpighi, P. Régnault's *Lettre d'un physicien sur la philosophie de Newton*, Fontenelle's *Mondes*, and Voltaire's *Éléments de la philosophie de Newton*. Under Cramer he also studied the famous *Logique de Port-Royal*: it interested him greatly, especially the sections on "the practical", which he could apply to his observations and experiments. At the "*Auditoire de Philosophie*", Bonnet became an assiduous student. Besides the public courses, he got private tutoring from both Calandrini and Cramer, in physics and speculative philosophy from the former, and in geometry and algebra from the latter. He would probably have learned more geometry and algebra, he concedes, had he not been so taken with natural history, but adds: "At least I received the germ of the geometric spirit, which Fontenelle said was more precious than geometry itself. The spirit of observation and analysis that developed in me later does not differ from the geometric spirit."[11]

But the direction of Bonnet's development while attending the "*Auditoire de Philosophie*" was probably influenced most decisively by an individual who was not even there, one he never met in person and would know only through the exchange of letters – Réaumur. Abbé Pluche had excited Bonnet's curiosity on the subject of insects, but when he saw Réaumur's *Mémoires sur les insectes* lying on the writing table of an instructor, de la Rive, he looked through part of it and knew immediately that here was a work which, by its thoroughness, could satisfy his curiosity. He asked to borrow the book but was categorically refused: the book, he was told, was over his head. Undaunted, he went to the public library to borrow a copy, but even there he was turned away with the explanation that such books were not lent to youths. Bonnet continued to pester the librarian until he relented, and then rushed home to read the work. That reading proved critical for his interest in natural history. He began on his own to repeat some of Réaumur's observations. In the course of observing a nest of caterpillars, he noted certain discrepancies with the master's results and wrote him, thus entering into a correspondence that continued until Réaumur's death in 1760.

One day a student exhibiting ordinary capabilities and expectations, the next receiving encouragement and direction from Réaumur himself; all this made quite an impression on Bonnet's "nascent *amour-propre*". If he suddenly began to see himself as a naturalist there could hardly be anything unusual in that. Henceforth, "nothing seemed preferable to the pleasures that accompanied the study of nature and the glory reserved to discoveries. Alas! I did not forsee that I would achieve a portion of this glory at the price of one of life's greatest goods, that I would one day regret having observed too much."[12] Bonnet's observations continued over the next several years, the results of which he religiously related to Réaumur. Cramer oversaw experiments during the same period and recommended more books on natural history.

At the age of nineteen Bonnet passed his exams and left the "*Auditoire de Philosophie*", an event which brought about a rather difficult confrontation between him and his father. It was time to choose a career: Bonnet's father wanted him to study jurisprudence and become a magistrate like his grandfather. Such a career would clear the family name and cancel his father's failures. But, repelled by the study of law and drawn to natural history, Bonnet resisted; he drafted a list of the advantages and disadvantages of jurisprudence on the one hand and philosophy on the other, not even bothering to mention natural history specifically since its advantages would be too difficult to demonstrate to his father. The elder Bonnet was unmoved as the advantages tipped in favor of philosophy; he would consent only to his son's working simultaneously at jurisprudence and at his observations. Hence Bonnet entered the "*Auditoire de Droit*" in the Spring of 1739, embarking upon four years he calls "the most thankless of my history".[13]

Considering the encouragement that Bonnet was receiving from Réaumur and the lack of understanding from his father at this time, it seems likely that Réaumur, in the role of mentor reinforcing the inclinations and interests of the son, now began to stand *between* father and son and, in effect, displaced and inherited the father's authority. So far was Bonnet *père* from being able to comprehend his son's real interests – or at least their legitimacy – that Bonnet gave up in his efforts to convey them. Rather, it was Réaumur who understood, and in the coming year it would be Réaumur who would put Bonnet on the road to fame. In May 1740 Bonnet decided to repeat an experiment of Réaumur's that had ended inconclusively – an attempt to ascertain whether plant lice (aphids) are able to reproduce parthenogenetically. When he made the surprising discovery that they did, Réaumur read Bonnet's findings to the *Académie de Sciences* and overnight procured for him a distinction about which he had only dreamed. By this action Réaumur demonstrated a high degree of confidence in his young correspondent.

With success in the field of observation, Bonnet, who never had been comfortable with philosophy, now appeared to regard it with open if mild disdain. He regularly met with a small "Society" of friends interested in discussing such metaphysical problems as, What constitutes the essence of the faculties of our soul? – a subject they correlated with the reading of Locke's *Essay Concerning Human Understanding*. But Bonnet deplored the time lost on such subjects, assuring his friends that they would learn more standing for fifteen minutes before a microscope than they ever would discussing substance and attribute; and they, in turn, replied that he would never be more than a simple observer of insects and small animals, an accusation at which he did not take offense.

For the next three years Bonnet remained extremely active in his naturalist studies. During the autumn of 1740, most of his time was taken up with continued study of aphids. He wrote his cousin, Abraham Trembley, on the subject; in replying, Trembley wondered whether parthenogenesis was limited to a single generation or if it extended over several, which led Bonnet to follow aphid parthenogenesis through several generations. Then, in a letter of May 5, 1741, Trembley informed Bonnet of his own work on polyps, and in particular of his discovery that they are animals which, when sectioned, reproduce complete wholes from each piece. Bonnet, like most of his contemporaries, was dumbfounded by the news. "Everything that I knew in natural history seemed like almost nothing compared to the *polyp*. It overthrew all my ideas and put my head, so to speak, into combustion."[14] Extending the experiments that had brought Trembley such astonishing results with polyps, Bonnet became preoccupied, in 1741, with experiments on freshwater worms, cutting them in two, ten, twenty and more parts. He set up extensive tables to record the growth of the different parts, at first on the freshwater worms and then on earth worms as well, continuing these experiments, uninterrupted, for three full years. There were also experiments performed on the respiratory system of caterpillars (1742), and on tapeworms (1743); besides these experiments, he made numerous other observations between 1740 and 1743, and communicated all his results to Réaumur.

At this point, having neglected law for natural history and now fearful of failing the public examination, he concentrated his efforts there in the early part of 1743, passed before an indulgent group of examiners, and was graduated. That same year, at the age of 23, he was honored with admission into the Royal Society of London, before which his principal observations on insects had been presented. Both Réaumur and Cramer encouraged him to put together a general history of insects, but his eyesight was becoming steadily weaker and he soon realized that he was not up to the task. Instead, he drew together the observations he had already made into a *Traité d'insectologie*, which was printed in Paris in 1744 and published the following year.

In his *Mémoires*, Bonnet wrote that he still found the "Préface" to that work well-reasoned; he even expressed surprise at finding various traits there of the logic and metaphysics that would be developed in later writings. One "fertile truth" in particular first appeared there – the idea that every production of nature as well as every part of the universe is incorporated within the hierarchic and almost infinitely nuanced gradation of one immense chain. "We see", Bonnet wrote toward the end of the "Préface",

> this innumerable multitude of organized and unorganized bodies placed one above the other according to the degree of perfection or excellence that is in each. If the order does not appear equally continuous throughout, it is because our understanding is still so limited; the more chains or degrees we uncover the more it will increase. When no more remains to be uncovered our understanding will have attained its greatest perfection. But can this be expected on earth? Apparently it is only celestial intelligences that enjoy this advantage .... And if, as I believe, all these chains, whose number is nearly infinite, form only a single chain, which reunites all the possible orders of perfection, it must be conceded that one would be able to understand nothing of that greatest or most elevated chain.[15]

Contained within this passage are all the essential assumptions by which Bonnet derives his taxonomy of terrestial beings, and further – but it comes to the same thing – by which he determines the nature of being, accounts for the unity and harmony observable throughout the structure of organized beings, finds a means of ranking all beings, and finally, indicates how progress in science may be charted.

These issues will be taken up again in the following chapter. At this point I wish merely to raise the question of Bonnet's source for the chain of beings. Without doubt Bonnet became one of the leading eighteenth-century proponents of the idea, perhaps the foremost proponent from the field of natural philosophy: in Cuvier's *éloge* the chain of beings is one of the central ideas for which Bonnet is remembered, particularly his effort to make use of this ancient idea as a modern taxonomic principle.[16] But, in light of Bonnet's self-proclaimed impatience with all metaphysics during this period, it is rather surprising that he produced, in the "Préface" to his first published work, a quite adequate summary of the chain and its relation to a harmoniously interconnected universe. Bonnet would have it that the idea followed quite naturally from reflecting upon the results of his observations: "Insects thus led me, unaware, to the metaphysics for which I had shown until then so great a repugnance and which would soon so delight me."[17]

It may well be that the idea of a chain of beings was so common in the eighteenth century that any attempt to trace its source in Bonnet's thought would be impossible; it probably is not going too far to say that the notion of a connected and hierarchic nature performed the same function in the eighteenth century that evolution has since the nineteenth: both concepts overcame the apparent (chaotic) dispersion and isolation of individual beings, transforming disorder into order, discontinuity into continuity, the former working within a spatial framework and the latter within a temporal one.[18] Still, if the idea of a chain of beings was all that common – and, more specifically, seen as a workable principle for the construction of a system of classification – it seems unlikely that Bonnet, so many years later, would still take such pride in this "fertile" idea found in his earliest work, or that he would be so well remembered for it. And it seems even more unlikely that Bonnet's readers would have concerned themselves with the question of where he had gotten his idea. But some of them did. One contemporary, M. Baume, assumed that Bonnet had gotten it from Buffon's treatment of nature's gradations in the "Premier discours" to his *Histoire naturelle*. Bonnet denies this, and points out that his *Insectologie* actually preceded the first volume of Buffon's *Histoire* by several years.[19]

For a specific source one could point to Leibniz, especially since Bonnet himself, in his later *Contemplation de la nature* and *Palingénésie philosophique*, connects the idea of a graduated unity of the universe to Leibniz's principles of continuity and sufficient reason. The *Insectologie*, however, was written in 1743, and Bonnet first read Leibniz in 1747. Of course he could have heard Leibniz's views expressed by others. By the time Bonnet read the *Theodicy*, there is no doubt that he found confirmation therein for many of his beliefs.

Bonnet's concept of the chain of beings could, however, have received its formulation as the result of reading Pope's *Essay on Man*. While Bonnet does not bring up Pope's name specifically in connection with the idea of the chain of beings, of the very few references in his "Preface" to the *insectologie*, several are to the *Essay on Man*. At the very least, they show that Bonnet had read the poem and still had it very much in mind as he composed it.[20] The *Essay on Man* begins with the presupposition of a universal harmony and order that Pope then goes on to elucidate. "A mighty maze! but not without a plan," (I, 6) he writes, at once locating this plan in a universal chain.

Vast chain of being! which from God began,
Natures ethereal, human, angel, man,
Beast, bird, fish, insect, what no eye can see,
No glass can reach; from Infinite to thee,
From thee to nothing. (I, 237–41)

Nothing is foreign: parts relate to whole;
One all-extending, all-preserving soul

Connects each being, greatest with the least;
Made beast in aid of man, and man of beast;
All served, all serving: nothing stands alone;
The chain holds on, and where it ends, unknown. (III, 21–6)

There are also progressive degrees of perfection within the vast chain, according to increased intelligence:

How instinct varies in the groveling swine,
Compared, half-reasoning elephant, with thine!
'Twixt that, and reason, what a nice barrier;
For ever separate, yet for ever near! (I, 221–4)

This gradation of intelligence leads to something else found in the passage of Bonnet's quoted above: he speaks there of the limitations of man's mental powers, which result from his relative place in the chain. Pope makes the same point.

Of systems possible, if 'tis confessed
That wisdom infinite must form the best,
Where all must fall or not coherent be,
And all that rises, rise in due degree,
Then, in the scale of reasoning life, 'tis plain,
There must be, somewhere, such a rank as man:
And all the question (wrangle e're so long)
Is only this, if God has placed him wrong?
Respecting man, whatever wrong we call,
May must be right, as relative to all. (I, 43–52)

Then say not man's imperfect, Heaven in fault;
Say rather, man's as perfect as he ought:
His knowledge measured in his state and place;
His time a moment, and a point his space. (I, 69–72)

Pope's message is one of moral edification: man's relative place, his limited powers, are meant to teach us intellectual modesty. Bonnet's concern is more with how these definite limits to human knowledge affect the entire process of scientific inquiry. Aside from that, it does seem apparent that the concepts and images presented in the *Essay on Man* regarding the unity of nature, the scale of beings, or the intellectual hierarchy could have been adopted by Bonnet.

How did the public react to Bonnet's first book? With the exception of one general criticism, the *Insectologie* was well received: the author was reproached "for having insufficiently spared the discreet delicacy of the reader in treating the '*amours*' of aphids".[21] It was a criticism Bonnet swears he had not anticipated.

Completion of the manuscript of the *Insectologie* coincided with the advent of the greatest crisis of Bonnet's life. His health deteriorated in January 1744; he grew weak; his eyes, which he believed he had overworked during his microscopic observations of aphids, were causing him "pain of varying intensity with each change in the barometer".[22] By 1745 it was only with extreme fatigue and pain that he was able to read or write. He saw filaments and feared cataracts, having read that they were preceded by such filaments. He could not work. Studying insects or using the microscope had become impossible. Moreover, an infection in his mouth forced the eventual extraction of many of his teeth in this same period. For nearly two years he did little but vegetate. "I felt only my privations."[23] Depressed and bitter, he blamed himself, blamed his anxiety to achieve fame for his loss of sight. Illness strengthened his religious convictions, and led him to a "reflective resignation". "Nevertheless", he immediately adds, "I did not cease to bitterly deplore, from time to time, the abuse I had caused to my health and principally to my eyes".[24] "I had decided that I would resent myself my entire life for the blow I had so rashly sustained on so delicate an organ." The sight of his microscope henceforth "always aroused a painful sentiment in me".[25] At the time, Bonnet was only twenty-five.

Such remained his condition until well into 1746. Réaumur quite naturally deplored the state of Bonnet's health, regretting the loss of so promising an experimentalist. Actually, Bonnet's experimental work did not end; he could no longer use his microscope and ended his observations of insects, but already in 1746, impatient to do something, he began to experiment with moss. And in the spring of the following year, his interest turned to the leaves of plants. It should be emphasized that Bonnet did not abandon natural history in taking up psychology and philosophy; nor did he give up the belief expressed in the "Préface" to the *Insectologie* that observation is a tool which ought to be applied far more widely than to natural history alone.[26] The viewpoint that he maintained through his later years was that neither his psychological works nor his cosmological speculations upon nature were categorically separate from his purely experimental and observational works. Rather, they informed one another. The more we observe, the more connections our minds make and, consequently, the more informed our generalizations will be; conversely, informed speculations and generalizations improve our methods of observing, organizing, comparing and judging.

Philosophy that ignores science held no more interest for Bonnet than did science that ignores philosophy.

In 1747 Bonnet's eyes somewhat improved, although reading or writing rapidly strained them and brought a return of pain. The filaments remained, increasing in size through the years. Such was the condition of his eyes for the rest of his life. There would be no more microscopic observations, nor would he read or write much; henceforth, books were generally read to him, his written work most often dictated.

From those years of illness, enforced inactivity and self-reproach, Bonnet rebounded into a period of great activity and creativity. This was the result of his turning from pure observation to speculation, from experimentation to philosophy, a transition that began during his illness. As for what precipitated this shift of focus in someone who had so recently scoffed at philosophical problems, whether it was the illness itself or whether it would have occurred anyway, the most satisfactory reply seems to be that there were several reasons for Bonnet's new attitude, some directly or indirectly connected to his illness and others entirely unrelated.

Twice in his autobiography Bonnet writes that he came to metaphysics by way of insects: first, when he points to the ideas that appeared in the "Préface" to the *Insectologie*, claiming that they derived from his observations; secondly, in reference to a debate he entered into in the summer of 1745 with Cuentz, a metaphysician from Saint Gall, author of *Système nouveau sur la nature des êtres spirituels* and a staunch defender of epigenesis. The debate was over the question of "souls" in animals that could propagate by means of sectioning. In the *Insectologie* Bonnet had argued that the recent discoveries, made first by Trembley with polyps and extended through his own research on worms, supported the hypothesis of preexisting germs.[27] In cases such as polyps Bonnet argued that preexisting germs were scattered throughout their bodies, and the souls of these animals he located within the germs. Cuentz opposed the view of animal souls, looked for an explanation of the new discoveries in a purely mechanical formation of organized bodies rather than in organic preformation, and took offense at the vast proliferation of souls brought about by Bonnet's germ theory.

By means of this debate, Bonnet was very early initiated into the epigenesist-preformationist dispute; he was quickly led from observation "to the metaphysics of animal reproduction."[28] Reticent about defending his position too heatedly, Bonnet turned to Cramer, who encouraged him to defend his views and advised him on how to dodge Cuentz's objections. Through this exchange with Cuentz, Bonnet learned how immediately philosophical questions could bear upon his work in natural history – a field he had previously supposed to represent a superior alternative to philosophy – and his impatience with philosophy dissipated: "Thus it was insects which recover from sectioning that caused my first step towards this metaphysics I had so greatly despised in my adolescence and had so often slandered because I did not understand it."[29]

This new interest in philosophy soon brought Bonnet into disagreement, for different reasons, with the two leading influences on his life, Cramer and Réaumur. At a meeting of his small society in May 1747, the question of freedom was discussed, and it led Bonnet to reflect "on the sources of our soul's determinations".[30] His friends were better informed about psychology than he, and it was in large part his ignorance that led him to ponder the issue, and to put his musings on paper. The results startled him. During his illness, religion had been his greatest consolation; yet the principles that unfolded from his reflections challenged the morality of human actions. Bonnet argued that the soul is endowed with its own activity but that this power is indeterminate. What determines the soul to act (to employ its will) are the judgments it makes. But since its judgments follow from the soul's comparing its ideas, and since all its ideas are originally acquired by means of sensations from exterior objects – since all its determinations, from the first to the last, originate in the impressions of exterior objects, and since it is these determinations that, in effect, direct the activity of the soul – it follows that there is no point at which human actions can be rendered accountable. He sent his results to Cramer, who was in Paris at the time. Cramer responded with an attempt to put each determination of the soul back within its power so that its actions might once again be ascribable. But Bonnet was not entirely satisfied and continued the correspondence, each reply from Cramer raising new questions for Bonnet. Rather than the controversy finding a resolution it was temporarily put to rest, to be revived in the Spring of 1749.

At this same time Bonnet was writing Réaumur, informing him of his growing interest in philosophy. In June 1747 Réaumur replied, expressing his pleasure that Bonnet had overcome his boredom and depression, but at the same time doubting the value of metaphysics. He suggested that Bonnet's present condition offered a good opportunity for him to compare the various productions of nature he had observed, but only so long as he formed hypotheses that could lead to experiments confirming or denying them. At the moment, however, maintaining his empiricist purity was not Bonnet's foremost concern. "It was in 1744", he writes, "that I was divorced from microscope and scalpel; and it was it 1747 that, in order to compensate for the privation of the study of nature, I surrendered myself to philosophical meditations".[31] These meditations led him back to writing, even though his eyesight made writing difficult. To facilitate the process, he taught himself to arrange lengthy passages in his head and to retain them for days, or even weeks, without alteration until he could dictate them. In light of his condition, this ability proved immensely useful for the composition of all his later works. The *Essai analytique*, for example, was composed entirely in this manner, and in that work, the ability is referred to as evidence supporting his views on the mechanism of memory.[33]

Bonnet's penchant for philosophy was further stimulated in the Winter of 1748 when he read Leibniz's *Theodicy*, "something I have always regarded as one of the principal moments of my thinking life".[34] In Leibniz, Bonnet found discussed some of the very problems that had been disturbing him – particularly the question of human freedom – and an approach that was both complementary to and more sophisticated than his own. Reading the *Theodicy* thus "marvelously increased my field of vision".[35] Still, he was not equally sympathetic to all the ideas found there, some of which seemed virtually impossible to comprehend. "I was lost, for example, with the famous *monads*, with those absolutely simple beings which, acting *only ideally upon one another*, nevertheless produce all the phenomena of nature".[36] It was a view as strange to him as Malebranche's occasionalism. Accustomed to physical notions, he failed to understand how such beings could give him the idea of matter. Leibniz's metaphysics thus seemed "to annihilate every production of nature I had taken pleasure in observing: the physical world vanished before me .... It was only [the discussion of] *freedom* and *optimism* that I actually understood well in the *Theodicy*," and on these topics the views of Leibniz were so "consoling" that Bonnet immediately incorporated them.[37] He also found welcome confirmation there for his position on preexisting germs and preexisting souls. The book served as a catalyst for Bonnet's thought, giving "a new life to my speculations".[38] Finding that on his own he had come to a number of views held by this great philosopher apparently gave him more confidence in his "speculations", so far afield from insects. Scarcely had he finished the *Theodicy* than he began dictating a work entitled *Méditations sur l'univers*. More than 900 pages in manuscript, it was never published in its original form, but it did provide the basis for several of Bonnet's later works: the *Essai de psychologie, Considérations sur les corps organisés*, and *Contemplation de la nature*. The *Méditations* provided Bonnet an opportunity to gain experience in composition, and to organize the various subjects that interested him.[39] As the writing progressed, he quickly came to enjoy the creativity involved. "The human spirit never feels its existence more agreeably than when it produces out of its own depths or when it gives new forms to the ideas it has acquired."[40]

He was beginning to *prefer* the pleasures of philosophical meditation and composition to those he previously had found in the study of insects. Upon hearing of this, Réaumur did not hide his disappointment: "You seem to go too far when you appear to prefer [the pleasure of meditation] to that in which you would make use of your eyes."[41] He doubtless felt betrayed by Bonnet's announcement. Despite the latter's circumstances, their friendship had developed out of a mutual adherence to the consummate value of patient observation coupled with modest interpretation, the older man nurturing the younger's talents. From the beginning, it was the thoroughness of the descriptions in Réaumur's *Mémoires* that Bonnet had found so valuable. Réaumur, he writes, had made him into an observer. Against the complaints of Buffon and de Brosses that Réaumur was long-winded, Bonnet countered that his fastidious recounting of details and steps were just what he had needed to try his own hand at naturalist studies. Réaumur wanted above all to inspire the young to study nature, and by his thoroughness he helped them develop "this spirit of observation, which is the best logic" since it "is a ceaseless logic in action, always leading to new discoveries".[42] Bonnet had been an ideal pupil. Thus his inability to continue observing insects was one thing, but breaking faith with the primacy of observation was quite another.

But Bonnet had not forsaken observation. Not only did he continue to make observations and to keep up with the literature in the field of natural history, he argued that his method had not actually changed, but that the procedures which had proven successful in his naturalist studies were just as useful when applied to the examination of mental faculties. At the beginning of the *Essai analytique*, he expressed the belief that at least his approach to the subject was novel, precisely because "I have undertaken the study of man as I have studied plants and insects".[43] At the same time it is true that, while he did not reject the value of his early discipline, he began to see it from a broader perspective. In later discussions of methodology, he recognized the limits of pure observation, even for natural history itself. Observations must still be interpreted; the components, once analyzed, must still be synthesized; and both steps are equally important. Careful attention is only one aspect of observation. Involved also is the "art of perceiving relations".[44] This idea of perceiving relations denotes, in the broadest way, not so much Bonnet's changed methodology as his changed perspective, probably the result of reading Leibniz. If all of nature is harmonious, if everything is linked together, then all things are joined by various relations. At the interpretative level it is no longer upon objects and particular beings that the naturalist must fix his gaze if he is to gain a better understanding of nature, but upon those relations themselves; for it is they that reveal the place and the significance of individual beings within the scheme of things. "Even in studying nature briefly, one soon perceives that all its parts are closely linked through various affinities. It is the research of these affinities that must occupy anyone concerned with the natural sciences."[45]

Bonnet may never have abandoned his belief in the primary importance of observation, but this probably was not clear to Réaumur, who, at the least, was disappointed at Bonnet's new interests. Soon, Bonnet would confront Cramer even more directly. In a sense, then, Leibniz became Bonnet's new "oracle", although it was not simply Leibniz: Bonnet's illness, his new philosophical interests, the confirmation of ideas he had already proposed and defended were all involved in the expansion of his conceptual field during this period.

Cramer's views on the question of human freedom had not satisfied Bonnet: "I always regretted that the solutions of M. Cramer seemed to me more subtle than substantial."[46] In February 1749, he returned to this subject and wrote an *Essai sur la liberté*, developing arguments there much as he had in the earlier *Propositions*. Freedom is defined as the active force of the soul; but this force, in itself indeterminate, is motivated as the result of other faculties operating within the soul. The understanding, which variously makes ideas out of impressions, compares these ideas, and notes their relations and oppositions, is one source of determination; the affections, or the soul's preference for particular pleasures – due to temperament, cultural mores, education, habits and prejudices – provide another set of determinations. But as all our affections either are given or derive from acquired ideas, all thoughts and actions appear to be predictable.

Yet Bonnet now concludes that we are free. The "*sentiment intérieur* convinces you of your freedom: and this sentiment is above all contradictions".[47] We all are free, but not to the same extent. Here Bonnet shifts away from a subjective to an objective idea of freedom, closely connected to his idea of an enchained hierarchy. The concepts of virtue and vice, good and evil are reducible, he says, "to those of *order* and *disorder*"; and "Order is constituted by the relations that derive from the state of things."[48] The better we perceive these relations, the better we are able to act in a manner conforming to the true order of things. Perceiving relations is a relational process demanding observation and reflection, all the tools of science and philosophy. By developing our understanding, then, we achieve pleasure and happiness, we act more virtuously and we act with greater freedom. For we are free insofar as our actions are in harmony with the existing order.

As for the meting out of punishments and rewards, Bonnet remains unconcerned. He is struck instead by the need for innumerable levels of freedom and virtue if everything is truly graduated in the universe, by the need for a hierarchy of perfection so that no two individuals occupy the same echelon. "Hence this infinite variety that shines in the universe. Hence this admirable *gradation*, by virtue of which all the degrees of perfection have been filled."[49] The reasons for a cruel Nero or a wise Anthony are to be found "within the context of the general system". Thus the order itself tolerates and even requires degrees of imperfection, which has a dampening effect upon punitive impulses. Moreover, when considering the actions of the individual, Bonnet cannot retract his opinion that any momentary decision of the understanding (and subsequent "free" act) is really only the effect of an entire background of habits and circumstances. Is someone who has been badly educated (that is, poorly trained to employ his rational faculties) to be judged the same as another person who is more aware of the relations constituting order? For purposes of civil justice, Bonnet proposes a graduated ethics corresponding to his view of graduated levels of freedom: each individual should be treated according to his particular situation; he should be judged relative to the degree of intelligence, freedom and perfection to be found within that individual, and not relative to some abstract moral standard, which no longer has any bearing.[50] As for the dispensation of justice in the next life, Bonnet will reject the notion of a punitive Deity.[51] There are only rewards awaiting us, for every species will participate in a uniform shift upward along the chain of beings, but individually, reward may be measured by degrees of retardation or acceleration of advance in relation to the rest of one's species.

Bonnet read his essay on freedom to his small society, where it met with little dispute. The group noted parallels with s'Gravesande's *Introduction à la philosophie*, which Bonnet had read. But when he then submitted the essay to the *Société des Gens de Lettres* – a large and distinguished group made up of older and prominent figures of Geneva, including a number of Bonnet's previous professors – it generated a good deal of dispute. Several members thought his principles and their consequences dangerous. Cramer, a member of the society, came forward and read a response in the form of questions directed to Bonnet. Bonnet saw that form of response as evasion, but nevertheless remained silent, believing it "improper for the student to contend with his master".[52] The rest of the members, however, applauded Cramer's queries and insisted that Bonnet respond; his friends also encouraged refutation. So he prepared a reply, and it provoked an even livelier debate than had the original essay because the specific points of confrontation were better developed. Cramer's responses surprised and disappointed Bonnet; the maze of subtleties and distinctions Cramer drew between certainty, necessity, varieties of necessity and impossibility were drawn, Bonnet felt certain, simply in order to reconcile philosophy and theology to the greatest extent possible. But when he approached Cramer in private to find out his actual opinions, Cramer only repeated what he had said before the society, and asked Bonnet to reconsider his own position. Bonnet did – and changed scarcely a word when he included the *Essai sur la liberté* in his *Essai de psychologie*.

With his two writings on the subject of freedom Bonnet confirms "that rational philosophy had really taken possession of my being".[53] In 1751 he decided to detach a part of his *Méditations sur l'univers* and try to publish it as a separate text. It was the material on psychology that he wanted to publish. The reason for this he attributes to youthful impatience and vanity; he was anxious to publish a "metaphysical" work since he was entirely unknown in that field, whereas his capabilities had already been demonstrated in natural history. His experiences with the *Société des Gens de Lettres* apparently had bred caution or doubt, however, as to how his ideas on psychology would be received, because he published the *Essai de psychologie* anonymously, steadfastly refusing to acknowledge authorship for 35 years. His only explanation: "I did not want to be exposed to a multitude of questions or objections."[54] Published late in the summer of 1754, the book did not reach Geneva until 1755; the first published copy Bonnet saw was one he found in a bookstore.

The *Essai de psychologie* deals at length with the faculties of the soul, charting their development through the stages of childhood from the registration of the first sense impressions made upon the newborn infant (and even upon the unborn fetus) to the activation of reason in the young adult. Since the soul's substance, laws of operation and faculties are in themselves beyond the reach of actual examination, Bonnet's particular interest is with the physical activities of the body: the structure and operation of the nerve fibers command his attention. But since the soul clearly appears to be affected directly by the body, and even – at least in the observable *effects* of its activities – seems to operate according to laws similar to those governing matter, Bonnet seeks the mechanistic tendency of its operations, as well as the relation between the operation of the nerve fibers and the soul's activities regarding the production and reproduction of ideas. Language and its relation to the abstraction of ideas is examined, as is the role of habit in the formation of ideas, tastes, opinions, mores and prejudices. Finally, these considerations lead to thoughts on education and its power to modify, reshape, correct, tame and acculturate the raw results of habit. Annexed to the *Essai de psychologie*, and written in February 1754, is a very interesting essay entitled *Principes philosophiques*: it constitutes, in effect, a précis of the entire *Méditations sur l'univers*, condensing into one short work "my principal ideas on cosmology, psychology, theology, morality, physiology, natural history, etc."[55] "Of all my productions", Bonnet writes, "it is there that I have tried to bring together the most things in the fewest words".[56]

In his *Mémoires*, Bonnet's attitude towards these youthful works is genial, in part because they represent "the first *jouissance* of my reason", but also because they contain "the germs of my later writings".[57] But even before the publication of the *Essai de psychologie*, he was already dissatisfied sufficiently with the sketchiness of his analysis to be planning a second psychological work. A different approach was needed, one that would better present his principles concerning "the mechanism of our being" than had the *Essai de psychologie*, with its examination of the effect of sensory impressions at the various stages of human development. While on a walk in October 1754 he had an inspired thought: why not create an imaginary model, a human statue that could be animated at will? This would greatly simplify matters for the purpose of analysis. Immediately he began composing in his head the entire "Introduction" and first chapter. He would examine the effects of sensations on each of five senses, beginning with sight. This was as far as he had gotten when he had the pleasure of chancing upon his old tutor, Laget, who, upon learning the nature of the work Bonnet was planning, pointed out to him that Condillac had already published a book using the idea of a statue. "As he spoke", says Bonnet, "a sensation of coldness spread through my veins".[58] He had not read Condillac's book, but resolved immediately to abandon his own. Laget assured him that there was no need to do this, however; he had read Condillac and thought that Bonnet was more interested in analyzing the physical aspects of our constitution, and that he, moreover, could be much more thorough than Condillac had been. Bonnet read Condillac and agreed. Following Condillac, he did, switch from the sense of sight to that of smell, "it being the least complex sense; and this consideration appeared to me to be well within the spirit of the analytic method".[59]

The *Essai analytique sur les facultés de l'âme* continued to occupy Bonnet until 1759, but not without interruption. In 1755 he met and began courting Jeanne-Marie de la Rive; they were married May 30, 1756. It was not an impassioned romance: "If you can imagine a middle ground between love and friendship, this will give you the situation of my heart."[60] And that was sufficient. His infirmities worried him, causing him to judge himself "perhaps too severely".[61] She was equally worried by her own delicate health, afraid she might be a burden. As for the marriage proposal, Bonnet liked to recall that "I did not speak the language of a lover: it was not made for her or for me".[62] All in all, a marriage well-attuned to the ideals of the age: passions comfortably restrained, reason in command. In Bonnet's psychology, one will find very little interest paid the passions; and as for imagination, it never fares well beside attentive and scrupulous observation and analysis. A thousand small steps to "genius" are as effective as a single bound – and more sure-footed.

The de la Rives owned property in Genthod on the north shore of Lake Léman, and in June of the following year the Bonnets moved there. But an accident with lasting consequences soon disturbed the tranquillity of the new setting. While on a drive one day in the surrounding countryside, a wheel broke on the carriage, and they took a spill. Mme. Bonnet, pregnant at the time, miscarried as a result of the accident, and never entirely recovered: not only was she no longer able to bear children, but she was often ill and depressed over the next several years. Bonnet faced this difficult period in a manner reminiscent of his earlier crisis, resigning to God's will what could not be controlled and retreating into his thoughts, into "meditation", for solace and distraction. If his life's work had changed as a result of the former crisis, his present difficulties only reconfirmed the direction in which he had been moving. Previously he had looked first to religion for comfort; now, he immediately resumed work on his *Essai analytique*, which had lain untouched since his courtship and marriage. By October 1758, most of this lengthy work, which he calls "the greatest effort of my reason",[63] was already finished, and he sent it to his publisher, Allemand. It was completed by August 1759 and published the following year.

The *Essai analytique* created something of a sensation, receiving praise from philosophers and journalists for its manner of explicating the "economy of our being".[64] Two letters, included by Bonnet in his *Mémoires*, get at several of the central questions, objections and subjects of interest raised by the work. The first was from Gaubius, a student of Boerhaave and teacher of pathology and chemistry at the University of Leyden. It is filled with admiration for Bonnet's "mathematical method", by which he means the way that Bonnet was able successfully to induce a seemingly "mathematical rigor" on "so delicate a subject". Gaubius felt that the *Essai analytique* demonstrated the immateriality of the soul more successfully than had all previous arguments; that despite similarities to La Mettrie's *L'homme machine*, the mind is convincingly shown to be not passive but to contain an active principle properly called thinking; and that it therefore shows how the mechanism of the soul's operations can be made to favor a system *opposing* materialism. Not surprisingly, this analysis pleased Bonnet since it caught his intent: far more often, to his annoyance, he was taken to be a thinly disguised materialist. His argument supporting the notion of an immaterial soul is an argument from its simplicity: one can, he holds, through the *sentiment intérieur*, sense quite clearly and directly that, in the midst of all the organs, senses and sensations of the body, as well as all the faculties and activities of the mind, the soul is one, individual, *moi*, its own unique personality. This simplicity, pervading and encompassing the multiplicity of the mind's operations, this simultaneous identity and multiplicity, defies the mechanical laws of matter.

The second letter, from Léonard Euler, a mathematician of renown who also busied himself in such areas as astronomy, chemistry, physics and metaphysics, offered a critique of what he held to be the basic principle of the *Essai analytique*: the claim that each sensation is allocated to a specific nerve fiber. If true, the rest would follow "geometrically". But, he asks, is it true: Bonnet proposes but never proves that each time the scent of a rose reaches the nose of the statue, the same nerve fiber or fiber bundle is affected; the scent of a violet, he assumes, again without proof, acts on other nerves. How can he refute a counterclaim that both odors work on the same nerves in different ways to produce different sensations? Knowledge of this subject, Euler points out, is very limited, particularly regarding the sense of smell. And sight, where it is better understood, is of no assistance to Bonnet's argument since, in the case of the optic nerve, the same fiber appears to represent different objects.

Bonnet had proposed his special fiber theory for a particular reason: he was trying to explain the mechanism of memory. Why a "mechanism"? Since illness and accidents can destroy memory, he determined that the basis of memory must be material, that it is not an aspect of the soul's operations. But how could sensations and ideas be stored by nerve fibers that produced different sensations every time they were moved differently? These fibers would have to retain their determination for a long time in order for the remembrance of particular sensations to be transmitted to the soul; and this requirement suggested that particular sensory impressions attached themselves to particular nerve fibers. This specific fiber theory was important to Bonnet's argument because, having located memory in the body, he felt compelled to explain how it could function in this material foundation. That is why, in order to come to an understanding of the mechanism of sensation, "It appears to me that it will always be necessary to return to whatever it is that yields the phenomena of memory".[65] Still Bonnet could not agree with Euler's claim that the principle of the diversity of sensations is the foundation of the *Essai analytique's* entire psychological edifice, and that if it is correct then all the rest is "geometrically demonstrated". He felt there were so many conjectures and unrelated subjects in the work that such a statement did not make sense. Finally, Bonnet disclaims having divined the true mechanism of the reproduction of ideas and says that he attempted "only to show how it would be possible for all the acts of our soul to correspond constantly to certain natural or acquired movements and determinations of nerve fibers".[66]

Bonnet returned to his *Méditations* manuscript after finishing the *Essai analytique*, this time intent upon extracting the part concerned with generation. He accomplished this task in February 1762, but not without having added a considerable amount of new material: the extractions from the *Méditations* comprised the first eight chapters; the rest was all new. In this book, which he called *Considérations sur les corps organisés*, his favorite hypothesis was set out and defended – that organic wholes, far from beginning their formation at the moment of fertilization, preexist in miniature in the egg of the female. Since the structure preexists from the original creation of the world, questions of generative formation are replaced by a close scrutiny of organic *development*. The all-important questions aim at resolving the "problem of growth", and central to this concern is the place of "nutrition", from conception on through all the stages of development: in plants and animals alike, the rapid ulterior development of the performed germ following fertilization is seen as the result of contact with the stimulating (irritable) and nutritional qualities of seminal fluid.

Bonnet became one of the leading proponents of preformation theory; and between this hypothesis and the opposing epigenetic systems, which sought formative forces to explain generation from the time of conception, contemporary scientific evidence seemed to confirms the validity of preformation.[67] Originally Bonnet had been disinclined to publish the material on organic reproduction in the *Méditations* because none of the facts upon which his principle of preexisting germs rested was "as decisive as I wished."[68] But studies by his friend Albrecht von Haller made on chicken eggs during incubation, the results of which he reported to Bonnet in the Summer of 1757 and then published in *Sur la formation du coeur dans le poulet* (1758), appeared to provide just the sort of confirming evidence Bonnet had been waiting for. In particular, Haller noted such continuity between the intestine of the chick and the membranes of the yolk that it appeared to establish convincingly that the "germ" belonged to the female and that it preexisted in the egg. On the basis of his observations, Haller cautioned against the common tendency to believe that the formation of organized bodies originates with the perceptibility of their organic parts. "One fails to consider that the repose, the smallness and the transparency of some of these parts can render them invisible to us, even though they actually exist."[69] That the structure could preexist its appearance was important to the preformation hypothesis; this analysis pleased Bonnet immensely since it coincided with his own ideas on generation in the preformed germ. In October 1758, he sent the relevant sections of his original *Méditations* to Haller for his judgment, and Haller urged him to publish them.

Haller, an epigenesist since news of the polyp, returned to a preformationist position and first stated his reconversion in his letter to Bonnet of September 1757.[70] Certainly one reason for the shift had to do with his work on the egg from 1755–57. But apart from those experiments Haller had grown increasingly dissatisfied with the logic of the forces on which epigenetic arguments relied to accomplish the task of organization. Recent studies of Haller have emphasized the importance of this issue to him, particularly in his reaction to Buffon's *moule intérieur*. How, Haller finally finds it necessary to insist, can one simple force accomplish such complex results? "I do not find in all of nature the force that would be sufficiently wise to join together the single parts of the millions on millions of vessels, nerves, fibers, and bones of a body according to an eternal plan . . . M. Buffon has here the necessity of a force which seeks, which chooses, which has a purpose, which against all of the laws of blind combination always and infallibly casts the same throw."[71] Thus the evidence of his experiments and the questionableness of the organizing capacities assigned to vague forces (his complaint against Wolff's *vis essentialis* was the same)[72] combined to return Haller to the preformationist fold.

The *Considérations* was well received. Much to Bonnet's irritation, however, Malesherbes interdicted it in France on the grounds that it was a metaphysical work possibly dangerous to the public. Undoubtedly, preformation did offer the possibility of such an interpretation, but Bonnet denied that his hypothesis had anything to do with metaphysics, claiming that he had tried simply to demonstrate that so-called "generation" creates nothing but is only the development of an already formed germ.

What remained of the original *Méditations* was a section that connected natural history to the harmonious enchained order of the universe. Bonnet had doubts about the strength of this material. Not only was it more speculative; parts of it would perhaps be better characterized as effusive reverie. Already a distinguished author, and one who prized his recent publications for their seriousness, it was with a mixture of "parental love" and embarrassment that he approached this segment of his youthful creation. Still, he was one to toss what he had written on the fire. More importantly, the material, which dealt with universal harmony and with the chain of beings, was not simply one more part of the original manuscript; it was its motif. This can hardly be doubted if one only looks at the precis of the *Méditations*, the *Principes philosophiques*. And there is even more direct evidence of this: after referring to the *Méditations*, Bonnet at one point immediately adds, "which I have since entitled *Contemplation de la nature*", the title of the book on which he worked in 1763 and 1764 and published in Amsterdam in the Fall of 1764. Although the third book from the original text, Bonnet pronounced it to be the original *Méditations* retitled, for it contained what he took to be the main subject of the earlier work. As with the other extractions from the *Méditations*, Bonnet enlarged the scope of the original material. The *Contemplation* begins with a lengthy consideration of the chain of beings, but then takes up the physiology of plants and animals as well as various issues of natural history.

The work went through several editions and translations. Spallanzani used it as a text in his classes at the University of Pavia. Its success surprised Bonnet. Having already published "two other works that had greatly exercised the strength of my understanding and in which I had attempted to probe some very difficult subjects",[74] the *Contemplation*, by comparison, seemed superficial. But he saw that it reached a different audience from that of his earlier works, readers with a general curiosity about the working of nature "who would have been less well served by a more erudite work".[75] That analysis does not seem entirely accurate, for, as I have already pointed out, it was Bonnet's fairly thorough effort in the *Contemplation* to substantiate the principle of continuity, through an elaboration of the chain of beings as a taxonomic principle, that gave an overarching unity and appeal to his survey of natural history.

The last of Bonnet's principal works, *Palingénésie philosophique, ou idées sur l'état passé et sur l'état futur des êtres vivants*, was begun in March 1768 with the idea in mind of a short essay of 20 or 25 pages, an essay that would return to an earlier discussion in the *Essai analytique* in which Bonnet had pondered the future rebirth of man, this time in order to include all living beings in those reflections – "to render probable to the eyes of reason the survival of animals and the perfection of all their faculties in a future state".[76] And not just animals: plants, too, are considered likely candidates for such a future restitution. This original idea was not abandoned; but before the text was finished – and it was composed very rapidly over the next fourteen months – Bonnet had once again worked through all his views on the "economy" of humans and animals, on the origin, growth and reproduction of organized beings, on the universe, on God. The *Palingénésie* had, in effect, little by little become a supplement to his three earlier works, with the inclusion in its last sections of what he thought to be the best proofs of Christianity.[77]

A future for animals and plants rests on the proposition that all living beings have souls. The question is whether all their operations can be explained mechanically (this Bonnet does not deny, but says he finds all such explanations more or less forced), or whether their actions suggest the presence of souls. Bonnet supports the latter view and seeks confirmation on the basis of analogies, first between the organs of animals and those of humans, which takes the *Palingénésie* into lengthy reviews of physiology, and secondly, between animal and human psychology. Regarding the latter comparison, Bonnet applies views developed in his works on psychology. Animals have a will, but it is directed only by sensations. They lack intelligence. While they have the capacity to retain these "sensible ideas", to compare them and derive judgments, they are incapable of raising them to the level of abstract ideas. Their situation, all things considered, is analogous to that of a child before it becomes a thinking being – through the development of its organs, and through the acquisition of language and education. Hence, "Animals today are in a state of infancy",[78] and will perhaps achieve, in future life, what humans are already capable of through physiological development (faculties of reflective thought, which blossom at a specific stage of growth) and cultural heritage ("gift" of language, education). Plants are rather further down the scale, both in their organizational complexity and in their mental capabilities, but this is no reason to deny them a soul. Coinciding with these efforts, in the *Palingénésie*, to demonstrate by means of analogy, several long sections of the *Contemplation* also are given over to showing existing parallels between animals and plants.[79]

Central to Bonnet's speculations on a future state is his determination to give the resurrection message of scripture a "physical" interpretation. Certainly he is motivated throughout by a desire to assure himself of the reality of resurrection; his attempt to extend "survival" to animals is, he confesses, only to make a future life more certain in the case of man.[80] But evidence for the transition into this future state is sought in the present make-up of living beings; the transformation is to be accounted for entirely by natural means, without divine intervention. Recalling his principle that man is the union of two substances, and therefore as much physical as spiritual *in essence*, Bonnet reasons that the dogma of resurrection has thus far been misrepresented as involving only the soul's immortality, when it was the immortality of man in his entirety – man as a "mixed" being – that concerned the Evangelists. This "physical" interpretation fulfills a deeper need: adherence to the principle of continuity. A "natural" passage from past to future means, in every way, fewer dichotomies between the present and the future than we are led to believe from Christian eschatology. We reappear on the same earth, and although our surroundings as well as our bodies have undergone great changes, although new relations are everywhere at play, there remains an uninterrupted connecting link: our past memories are borne forward, intact, in yet another performed "germ" located in the brain. This germ is the physical aspect that never perishes, that unfolds in the future according to the laws of germ development, that accounts for the physical base of memory and that maintains the continuity of personality across death's lacuna.

Under Bonnet's guiding hand the future state no longer represents difference but an ongoing progression from the present so that "the present life is the first rung of a chain lost in eternity".[81] And if all this applies equally to animals, if there is no reason to suppose a radical bifurcation between the future paths of human and other living beings, then continuity enters once again in support of Bonnet's suppositions, hopes and dreams. Before he is finished Bonnet moves through physiology, natural history, comparisons of animal industry and human psychology, a theory of cataclysmic earthly upheavals, and interpretations of Christian eschatology – looking in all these areas for evidence to support animal survival, seeking "the links uniting their present and future economies which are two economies in a single or continuous chain", and rejecting, as an impoverishment to the universe, "the common opinion which condemns all organized beings with the exception of man to eternal death .... My view preserves all these beings and gives them permanence".[82] The principle of continuity itself derives from the principle of sufficient reason, and there is scarcely ever a sufficient reason for discontinuity.

But at what cost is all life redeemed and resurrection naturalized? Some read the *Palingénésie* as actually dispensing with rather than clarifying the dogma of resurrection, for, in effect, Bonnet does away with the notion of one temporal life on earth at the present to be followed by an immortal (atemporal, supraterrestrial) life. In accordance with his desire for a sequential continuum to match the spatial continuum and the hierarchy of the chain of beings, a number of substitutions are made in Christian doctrine. The radical dichotomies (terrestrial/supraterrestrial; mortal/immortal) are replaced with innumerable, serial, terrestrial lives. All supernatural intrusion has been carefully dispensed with: biblical accounts of the Deluge and even of the Creation are read as records of revolutions within recorded history, just as biblical accounts of the end of the world are read as no more than a description of the next (but not the last) revolution. In short, the terrifying discontinuities of eschatology are naturalized, terrestrialized – tamed – and brought within the bounds of continuity. All this was comforting to Bonnet, as well as to some readers of the *Palingénésie*, but not to everyone. M. de Maupeou, then Chancellor of France, for a time had the book interdicted.[83]

Humans, and perhaps all living things, form an uncompromising, although mysterious, union of material and spiritual substances. This dualism confronted Bonnet once again as he began, at the time he was finishing the *Essai analytique*, to ponder writing a "*Morale philosophique*". The extent of the plan eventually scared him off: first, to demonstrate the nonarbitrariness of moral principles, would have required the derivation of these rules "from the relations that enchain all beings"; and secondly, since the complex of relations adjoining each being to the harmonious whole followed of necessity from the nature of those beings, morality would have to give equal attention to both substances: "morality must therefore be occupied with the preservation and perfection of material substance just as it is with [the perfection of] immaterial substance".[84] Moral principles would have to encompass all questions concerning the preservation of health and life; then again, they would have to encompass questions concerning the immaterial substance, which would lead into psychology, logic and natural law. Although it was never written, Bonnet did write what was to have been the "Introduction" to this work in 1767, and finally published it in the collected edition of his works in 1783 as the *Philalethe*.

What Bonnet's plan did not require was the intervention of the Supreme Legislator handing down moral dictates, for all moral principles are deductible from the nature of things and from the relations existing between these things. His *Morale* had to be a practical guide for Christian and atheist alike. "It is through a consideration of the faculties of my soul in their practical application that I acquire the philosophical notion of the moral being, and through it of the moral order."[85] Simple daily experience is the "practical application", from which we learn that only certain "exercises" of our faculties are life-preserving or life-enhancing and therefore related to our well-being. To exemplify his point, Bonnet described how temperance becomes a part of our moral character: "This organized body that makes up so essential a part of my being and to which my soul is united . . . would not be preserved without the daily introduction of foreign matter into its interior to replace that which intestinal movements dispel . . . . Between the action of these organs and the foreign matter on which their action is deployed relations exist such that the immediate result is the incorporation of this matter into my own substance. The result is a law of my being; but of my purely physical being." Physical laws determine the amount of food we should eat, and we only need overeat to discover just how binding these laws are. Overeating, perhaps a momentary bliss (an apparent good), leads to actual agony. "My reason thus deduces certain consequences from my physical constitution"; and it is due to the entrance of reason that morality is involved. Humans, endowed with reason, are able to deduce the laws of their physical being from their experiences, laws which must be obeyed for the sake of their well-being. The consequences deduced from overeating therefore become a law of one's being too, but of one's moral rather than physical being. "I call them *moral laws* because I discover them only with the aid of my reason, and because they govern only beings endowed with reason. Hence, in this particular case, *temperance* becomes a law of my moral being."[86]

"The physical corresponds to the moral, and the moral to the physical", Bonnet had written long before in his *Essai sur la liberté*. And why is this so? Because everything in the universe is connected within this all-encompassing order. Thus, "The ideas of just and unjust, of virtue and vice, of good and evil are reduced to those of order and disorder." Human beings are in the unique position of consciously knowing this order, at least to some degree, and "The *virtuous* man is the one who conforms to the order." "Thus morality consists essentially in the conformity of man's actions and judgments with the established order, or what comes to the same thing, with the state of things."[87] More specifically, "To act in a manner conforming to order is to act in a manner conforming to the relations that we observe between things."[88]

From overindulgence we acquire knowledge of the relations existing between the need for food and the organs of digestion. If we reflect on our fellow beings, we discover "relations of dependence" designated by the needs of our nature. "We will also notice that these needs are reciprocal, and that they join together all the members of humanity." From these "relations of dependence" we therefore discover the "laws of sociability". But it is not only with mankind that we maintain relations: "Through need, through pleasure or through knowledge I find that I am connected with everything."[89] These, then, are the general considerations that lead to the notion of a *moral being*: intelligent beings conforming their actions to the order they gradually discover through the various relations which they sustain with their own bodies, with their fellow beings and with the whole of their environment. It follows that the more intelligent the individual is, the greater his moral possibilities and responsibility (in sum, perfection) will be, for he is in a better position to know and to observe nature's order.

Such considerations lead to questions of how we know, how our intelligence operates, how reliable human knowledge is; and they lead to other questions concerning our needs, interests and pleasures: that is, to questions of psychology. To answer the questions raised regarding moral man we are driven back, in the end, into psychology and into physiology. A correspondence exists between the physical and the moral, and the same holds true for psychology and physiology. Humanity, that elusive conjunction of intelligence and matter, always requires simultaneous analysis at both levels. Thoughts that arrive through the body; ideas that are stored in a material depository; passions and discomforts, pleasures and illnesses that are all passed back and forth through this invisible gate and out into the adjoining world: if all this is man, how could Bonnet discover even so much as a change of direction in his work as he moved from the great questions of generation to those of abstract thought, and back again? "Psychology and physiology enlighten one another; they contain many common aspects since man is the principal object of both. Now if everything is closely connected in man, if he is a marvelous system of relations, then the sciences concerned with man are necessarily enchained to one another. Without doubt, it is as a natural consequence of this intimate connection that I have been led to meditate successively on two of nature's greatest mysteries, the mechanism of the mind's operations and the origin of beings."[90]

What about the mind's operations? In pursuing them, how far do we ever really move away from the body, from their effects upon the body, from the laws governing matter and from the methods appropriate to unveiling those mechanical laws? Not very far. "The science of the soul", Bonnet insists, "like that of the body, rests equally on observation and experimentation".[91] We are ill-equipped intellectually to do more than study the material and mechanical aspects of human psychology; "in truth, what can be said of the soul considered in itself? We know so little about it!"[92] Certainly we are free to speculate, and Bonnet is quite willing to engage in this too; but so far as a "science of the soul" is concerned, we are forced back into physiology, limited to analyzing the soul's effects, the mechanical operations produced upon the body. Thus, in his psychology, Bonnet concentrates upon the physical – with his analysis of nerve fibers and his analysis of the functions of a physical memory – rather than upon the metaphysical, believing that observation and analysis, the methods learned in his youth, can apply here too, seeing the task of psychological investigation as not differing essentially from any other investigation of the human body. One studies relations, one looks for the laws resulting from them. The task remains the same. To uncover man's place within his environment, his place in the order of things. At bottom it is a moral endeavor, for the more one knows of these things, the better one will be able to adapt himself to the order.

Psychology continued to occupy Bonnet up until his death in 1793, at the age of 73. The publication of Kant's *Critique of Pure Reason* presented Bonnet, in his later years, with a challenge to the "psychological principles" with which he had worked all his life, regarding the way in which we come to knowledge of the exterior world through sensations (experience),[93] as well as the way in which we have access to a direct apperception of our soul, of its unity or simplicity, through the *sentiment intime*.[94] At the end of 1788, Bonnet was moved to write several precis and commentaries on that work, and to conclude that Kant's strange doctrines tend to reverse all received opinion in rational philosophy, "and even to annihilate all metaphysics. He ceaselessly shocks the sources of all our knowledge, *sentiment intime* and experience; and he gives no attention to the natural generation of our ideas, to the subordination of our faculties, or to the manner in which they develop through one another. His science is *terminological* rather than *real*, and useful only in encumbering young men with subtleties they cannot unravel."[95] Bonnet predicts a short reign for Kant's philosophy; but proper and complete refutation – Kant's being a "terminological" science – would demand the compilation of a dictionary of his terms.

Certainly Kant reversed the direction Bonnet had moved in, for the latter's efforts to establish a "science of the soul" had meant, from the start, binding psychology more directly to physiology".[96]

In this area it was, in particular, the problems raised by the physiological foundation of memory that still absorbed him. In 1786–7 he wrote a number of essays on reminiscence which were critiqued by Trembley, among others, to whom they had been sent for that purpose. These papers, collected together in his unpublished manuscripts under the title *Sur la reminiscence*, actually show him encountering more and more difficulty in attempting to bring the mechanism of a physiological memory into agreement with the laws of motion.[97] And they reveal something else: a spirit that runs through all of Bonnet's work, manifesting itself in greater concern always for the difficulty of the problems than for finding vindications of solutions he had already posed. A partisan of the new methods of observation and experimentation, Bonnet had learned to cultivate patience and perseverance and to avoid dogmatism. As he wrote in a letter to Spallanzani on January 17, 1771: "I have always sincerely sought the truth and I have warned a hundred times that I never flatter myself with the thought that I have always found it. So don't spare me my errors, and criticize me wherever you judge it necessary . . . . When nature pronounces against me you must not indulge in the language of friendship; and I will be the first to submit to its decisions."[98]

NOTES

[1] *Mémoires autobiographiques de Charles Bonnet de Genève*, ed. Raymond Savioz (Paris, 1948), 40. Hereafter cited as *Mémoires*. Translations from French texts are mine throughout.

[2] He took only one trip in his lifetime when he visited his friend Haller at Roche, a small borough located at the east end of Lake Léman. Bonnet lived on the north side. See *ibid*., 381.

[3] *Ibid*., 290.

[4] *Ibid*., 180. The biographical information presented here is in no way meant to be critical or definitive; its purpose, rather, is simply to provide the reader with some sense of Bonnet's circumstances and chronology. Raymond Savioz, in his *La Philosophie de Charles Bonnet de Genève* (Paris, 1948), in his "Introduction" to the *Mémoires*, and in his editorial notes found at the end of the latter book, presents additional information, and rounds out the biographical material with details and speculations on traditions, influences, forerunners and followers, most of which have, happily, taken up little space here. As for biographical details, Savioz takes nearly all of them from Bonnet's *Mémoires* – and I simply turned to that source to construct my own picture. The weighting of the various elements involved in Bonnet's shift to psychology and philosophy, the theory of the mentor role of Réaumur and Cramer and its consequent significance in Bonnet's development, these are my own reconstructions based on my reading of Bonnet's *Mémoires* and I trust that I have included enough supporting material to substantiate them.

[5] *Mémoires*, 40.
[6] *Ibid.*, 42.
[7] *Ibid.*, 46.
[8] *Ibid.*
[9] *Ibid.*
[10] *Ibid.*, 45, where Bonnet quotes from d'Alembert's *éloge* for Jean Bernoulli.
[11] *Ibid.*, 48.
[12] *Ibid.*, 52.
[13] *Ibid.*, 59.
[14] *Ibid.*, 65. See Abram Vartanian, "Trembley's Polyp, La Mettrie, and Eighteenth Century Materialism", *Journal of the History of Ideas*, XI: 3, 259–86.
[15] Quoted in *Mémoires*, 81.
[16] In face of the reproach, concerning his scale of natural beings, that he had failed to provide an account of the exact placement of productions on the scale, a new addition "risked" inserting such a scale at the end of the "Préface". See Bonnet, *Traité d'insectologie* in *Oeuvres d'Histoire naturelle et de Philosophie*, 18 vols. (Neuchatel, 1779–83), I, xxi. For the scale inserted there, see below, Chapter II, 39–40. In the *Contemplation de la nature*, in *Oeuvres*, VII–IX, published about twenty years later, the scale was presented in greater detail.
[17] *Mémoires*, 81. See *Traité d'insectologie*, ix–xi for the same point.
[18] Cf. Michel Foucault, *The Order of Things* (New York: 1970), *passim* Chapters 3, 5, 7.
[19] *Mémoires*, 81–2.
[20] It seems probable that he had read it recently since he is so often reminded of it. So far as I know he does not refer again to Pope except for several citations in the *Palingénésie philosophique* (1767) in *Oeuvres*, XVI–XVII, Part XIII, Chapters V, VIII.
[21] *Mémoires*, 83.
[22] *Ibid.*, 84.
[23] *Ibid.*, 85.
[24] *Ibid.*
[25] *Ibid.*, 85.
[26] Repeated, among other places, in the "Préface" to the *Essai analytique sur les facultiés de l'âme* in *Oeuvres* XIII–XIV, and in the "Préface" added to the first volume of his collected works.
[27] *Mémoires*, 80.
[28] *Ibid.*, 92.
[29] *Ibid.* Again, on the following page, he writes that he "took up metaphysics in 1745 only through some of its most general relations to natural history, and in particular to insects".
[30] *Ibid.*, 93.
[31] *Ibid.*, 335.
[32] *Ibid.*, 335, 99, 40–1.
[33] *Ibid.*, 224; *Essai analytique*, § § 816, 817.
[34] *Mémoires*, 100.
[35] *Ibid.*
[36] *Ibid.*
[37] *Ibid.*, 101. For Leibniz on freedom see, e.g., *Theodicy*, part I, § § 34–52.

38 *Mémoires*, 101.
39 *Ibid*., 104, 208.
40 *Ibid*., 104.
41 *Ibid*., 105.
42 *Ibid*., 51, 208.
43 *Essai analytique*, vii.
44 Bonnet, *Considérations sur les corps organisés* in *Oeuvres*, V, Chapter IX, art. CLIII, 260.
45 Bonnet, *Tableau des considérations sur les corps organisés* in *Oeuvres*, XV, 73–4.
46 *Mémoires*, 109.
47 *Ibid*., 112.
48 *Ibid*.
49 Bonnet, *Essai sur la liberté*, § XLIX, quoted in *Mémoires*, 114.
50 *Mémoires*, 131.
51 In *Essai de psychologie* and in *Palingénésie philosophique*.
52 *Mémoires*, 117–8.
53 *Ibid*., 131.
54 *Ibid*., 168.
55 *Ibid*., 175.
56 *Ibid*., 168.
57 *Ibid*., 168, 175.
58 *Ibid*., 176–7.
59 *Ibid*., 177.
60 *Ibid*., 181.
61 *Ibid*., 180.
62 *Ibid*., 181.
63 *Ibid*., 356.
64 *Ibid*., 193.
65 *Ibid*., 199.
66 *Ibid*., 202. How the olfactory nerves – the focus of the *Essai analytique* – work seems to be an issue hardly less settled today than when Bonnet was writing, and one current theory sounds reminiscent of Bonnet's:

It is not known how the olfactory cells are fired by an odorant. According to one view, a hole is poked in the receptor membrane, launching depolarization, but other workers believe that the substance may become bound to the cells possessing specific receptors for it and then may just sit there, somehow displaying its signal from a distance, after the fashion of antigens on immune cells. Specific receptor proteins have been proposed, with different olfactory cells carrying specific receptors for different "primary" odors, but no one has yet succeeded in identifying the receptors or naming the "primary" odors.

Training the cells for olfactory sensing appears to be an everyday phenomenon. Repeated exposure of an animal to the same odorant, in small doses, leads to great enhancement of acuity, suggesting the possibility that new receptor sites are added to the cells.

This is taken from Lewis Thomas, *The Lives of a Cell* (New York, 1974), 38–9.
67 Bonnet directed himself in particular against Buffon and Needham.

68 *Mémoires*, 209.
69 Quoted in *ibid*.
70 See Shirley A. Roe, "The Development of Albrecht von Haller's Views on Embryology", *Journal of the History of Biology*, **8**, No. 2 (Fall, 1975), 174.
71 In Haller's preface to the German translation of the second volume of Buffon's *Histoire naturelle* (1752). Quoted in *ibid*., 177. See also Jacques Roger, *Les sciences de la vie dans la pensée française du XVIIIe siècle* (Paris: 1971), 705ff.
72 *Ibid*. Also see Shirley A. Roe, "Rationalism and Embryology: Caspar Friedrich Wolff's Theory of Epigenesis", *Journal of the History of Biology*, **12**, No. 1 (Spring, 1979): "why", asks Haller of Wolff, "does this essential force, which is unique, form the parts of the animal, which are so different, always in the same place and always according to the same model, if inorganic matter is susceptible to change and capable of taking all sorts of forms . . . . Nothing is assumed [by Wolff] other than an expanding and progressive force. I would expect nothing more from this than that the net of vessels would necessarily become larger in the future . . . . Why, at the site of this net, are formed a heart, a head, a brain, a kidney? Why in each animal is there an arrangement of its parts? To these questions no response is given." And to this challenge from Haller, crucial to Wolff's generation theory, Wolff had no answer. The quote is from *Elements physiologiae*, found in *ibid*., 23.
73 *Ibid*., 208.
74 *Ibid*., 220–1.
75 *Ibid*., 221.
76 *Ibid*., 248. See *Essai analytique*, Chapter XXIV.
77 This was also published separately as *Recherches sur les preuves du christianisme*.
78 *Palingénésie*, part VIII, Chapter IV.
79 *Contemplation*, parts X, XI. These parts were considered by Bonnet and his publisher to be among the most interesting of the entire work. For that evaluation, see *Mémoires*, 220.
80 *Ibid*., 238.
81 *Palingénésie*, part VIII, Chapter 1.
82 *Mémoires*, 248; *Palingénésie*, part VIII, Chapter V. See also *ibid*., part XIV, Chapters I and II.
83 See *Mémoires*, 250–1.
84 *Ibid*., 235.
85 *Philalethe* in *Oeuvres*, XVIII, Chapter VIII.
86 *Ibid*.
87 *Palingénésie*, part XV, Chapter VI.
88 *Essai sur la liberté*, § § LII–LIV, XLV, LX, in *Mémoires*, 114–5.
89 *Philalethe*, Chapter XIII.
90 *Tableau*, 61–2.
91 *Essai analytique*, xxvi.
92 *Ibid*., xxii.
93 Concerning experience, for example, Bonnet quotes from the *Critique of Pure Reason*: "Our knowledge derives from experience; but this knowledge is not of a nature satisfying to the understanding, which wants general truths, existing by themselves *a priori*." Bonnet replies: "If our knowledge 'derives from experience,' there are no *a priori* truths, and it is well-demonstrated in psychology that 'general truths' derive by

way of abstraction from sensible ideas." Again he quotes: "This *a priori* representation [of phenomena] is inherent in our nature, and since it is related immediately to the object, it is also intuition. It is thus a *pure a priori* intuition." And again he replies: "The author has said that 'our knowledge derives from experience.' Knowledge of an object thus derives from experience, and the representation of this object will therefore be unable to be a pure *a priori* intuition." See *Précis de la doctrine de Kant sur l'entendement pur*, Bibliothèque de Genève MS 759.

94 Kant also requires the absolute unity of the thinking being, without which it would be impossible even to say "I think"; but he makes the simplicity of the self a *logical* requirement (reason demonstrates its absolute necessity) rather than something known directly and intimately through some sort of reentry into ourselves, through a reflexive activity of the soul. See *Remarques sur quelques endroits du livre de M. Kant intitulé*: *Critique de l'entendement pur*, Bibliothèque de Genève MS 35–32 – 759.

95 *Précis de la doctrine de Kant*.

96 According to Savioz, in Germany Bonnet has been generally considered to be the initiator of physiological psychology. For an estimation of Bonnet's renown and "influence", see Savioz's *La Philosophie de Charles Bonnet*, 330ff.

97 See Trembley's "Objection contre l'opinion de l'Analyste sur la physique de la reminiscence" in *Sur la reminiscence*, Bibliothèque de Genève MS 35–34 –758.

98 Quoted in Savioz, *La Philosophie de Charles Bonnet*, 328–9.

CHAPTER II

TAXONOMY

Upon reading the *Contemplation de la nature*, one might be perplexed to find not the least sense of indignation on the part of this Christian naturalist as he describes the close proximity of man and ape. To the contrary, his ebullience if anything seems to flow over. While in the text one finds that "The ape is an outline of man: a rough sketch, an imperfect portrait but still resembling him", a note appended a number of years later goes much further: "What an enormous distance separates man from the dog. And yet, between man and dog, the chain is nearly continuous: and in reclimbing the length of this chain, the contemplator of nature arrives with surprise at a being which so resembles man that the features distinguishing it seem not so much specific characteristics as mere variations."[1] The analogy is brazenly drawn out: "What is one to make of a being which, while it is not properly man, yet has the stature, the carriage, the appendages and the strength of man"; that walks and sits like him, lacks a tail, has hair on his head and whiskers on his face, that has similar hands, feet, and nails; "that arms himself with stones and sticks in order to attack and defend himself; that is also ardent for females and for families of his species; finally, that is capable, through education, of discharging the services of an adroit *valet-de-chambre*, and of contracting the habits, manners and even an urbanity that would seem proper only to man?"[2] Bonnet speaks here of the orangoutang that had been exhibited in Paris.[3] The passage concludes by noting that resemblances extend beneath the surface as well, where its anatomy bears a marked similarity to man's, even to the extent that its brain has the same form and proportions.

How is this comparison, drawn out with such zeal, to be taken? If one sees in it an anticipation of the descent of man, he could not be more mistaken; it is precisely because Bonnet never considered such a possibility that he so avidly drew his analogy. He worked in a different perspective from that of nineteenth-century evolutionism: there, connections would be drawn in time, man would be linked with the ape by demolishing the cramped Christian conception of time and history, and by rebuilding a continuum backward by way of ancestral ties and chance mutations. In Bonnet's work, continuity resulted from a hierarchic and static arrangement of all the manifestations of nature linked vertically in a chain of beings until nothing – not a rock, tree, dog, or man – remained outside this all-embracing continuum: "Between the lowest and the most elevated degrees of corporeal or spiritual perfection there is an almost infinite number of intermediate degrees. The series of these degrees composes the *Universal Chain*. It unites all beings, links the whole world, embraces all spheres. A single Being alone is outside of this chain, and that is He who made it."[4] All things are ranked in a graduated scale of which we are able to see, and only imperfectly, the middle regions. And if we see breaks in the chain it is no doubt because we have misunderstood the natural order, for one thing is certain: "There are no breaks in nature; all is graduated, nuance. If between any two beings there were an empty space, what reason would there be for passing from one to the other? There is therefore no being above or below but what other beings are reconciled to it through some signs and distinguished from it by others."[5]

The ape is brought so near to man in order to leave no doubt that they touch within their adjacent space. Thus Bonnet prefaced his remarks on the orang by writing: "It is here especially that one cannot fail to recognize the graduated progression and find verification for the famous Platonic axiom that nature goes nowhere by leaps."[6] Moreover, if we are to address ourselves strictly to the continuity of nature, as perceived by Bonnet, we have far from completed the task of ordering when the orang has been ranked next to man; each ape and each man also must be arranged by imperceptible degrees along the scale of nature; divisions into species and classes are only rough approximations, no more than conceptual conveniences, and without basis in reality. "If nothing is severed in nature, it is evident that our classifications are not its own. Those that we form are purely nominal, and we should regard them as means relative to our needs and to the boundaries of our knowledge. Intelligences superior to ours will perhaps recognize between two individuals that we have placed in the same species more variety than we find between individuals of distant genera. Therefore these intelligences see in the scale of our world as many steps as there are individuals."[7]

If one sorts out the similarities and dissimilarities amidst the entire manifold of nature, what is to be made of this endeavor which became not just the preoccupation of one thinker, but an eighteenth-century phenomenon defining the very method of natural history itself? From where came this veritable mania for arranging, correlating, and classifying? It resulted in the broadest sense from the rationalistic preoccupation with unity. If the components of experience are not susceptible to strict unification, they are not classifiable either. By this postulate science and the concept of unity became interdependent: modern science founded itself on the sheer faith that the universe possessed uniformity, a notion that pervaded the eighteenth century.[8]

But how to distinguish the totality of that unity; how to break it up into comprehensible units and then reassemble it? It is in this matter of ordering that Michel Foucault finds the epistemological core of eighteenth-century thought. By his analysis, "nature throughout the Classical age did not exist in the first place as a 'theme,' as an 'idea,' as an endless source of knowledge, but as a homogeneous space of orderable identities and differences".[9] "[I]t was a question of establishing relations of identity or difference against the continuous background of similitudes."[10] As for natural history, it took observation as its methodological base and systematically began to restrict, not broaden, its area of experience; to look more selectively, indeed to give sight the almost exclusive privilege of arranging. Furthermore, as seeing became systematic, anatomy tended to lose its leading role; relations and resemblances in the eighteenth century were for the most part observed on the surface: "The plant and the animal are seen not so much in their organic unity as by the visible patterning of their organs. They are paws and hoofs, flowers and fruits, before being respiratory systems or internal liquids."[11] And there is a shared epistemological base, from the artificial System of Linnaeus, which selects a finite and relatively restricted group of characteristics, viz., sexual, to the natural Method of Adanson, which deduces characters step by step from total comparisons within empirically constituted groups. The common base "can be defined briefly by saying that, in Classical terms, a knowledge of empirical individuals can be acquired only from the continuous, ordered, and universal tabulation of all possible differences".[12]

This we see confirmed, really, in the "Premier Discours" to Buffon's *Histoire naturelle*, despite his famous polemic there against all systems, with special emphasis on that of Linnaeus. To recall his complaint, artificial systems, by selecting out and building around a single feature (whether, in the case of plants, it be the number of stamens, the configuration of leaves or flowers, or whatever), are guilty of arbitrariness, of falsely representing nature and, consequently, of being unscientific. "For who does not see that whatever proceeds in such a manner cannot be considered a science. It is at the very most only a convention, an arbitrary language, a means of mutual understanding. But no real cognizance of things can result from it." In judging the whole by a part, systems falsify through their one-sidedness. And should that part be tiny, as in the case of the stamens, one is forced into the posture of making identifications with a microscope in hand, meanwhile ignoring the whole range of readily observed similarities and differences constituting the true progression of nature. "The size, the form, the external appearance, the leaves, all the obvious features are useless for purposes of identification. Nothing is important except the stamens, and if one is unable to see the stamens, one can do nothing, one has seen nothing of significance. This large tree which you see is perhaps only a bloodwort." True method, then, must take into account all the features of the object under consideration. To do otherwise "is to voluntarily renounce the greatest number of advantages which nature offers us as a means of knowing her".

> And it appears to me that the sole means of constructing an instructive and natural system is to put together whatever is similar and to separate those things which differ. If the individual entities resemble each other exactly, or if the differences between them are so small that they can be perceived only with difficulty, such individuals will be of the same species. If the differences begin to be perceptible, while at the same time there are always many more similarities than differences, such individuals will be of different species, but of the same genus .... This is the systematic order which ought to be followed in arranging the productions of nature. Certainly the similarities and differences will be taken not from one feature only but from the whole organism.[13]

Linnaeus, for his part, did not deny the artificiality of his system, but there was no need to concede that it was therefore *arbitrary*, too, as Buffon claimed. If reproduction provided so important a classificatory device, "The endless fertility of this device arises from the Linnaean conception of natural order as a complicated network of formal similarities and differences. Any universal function, such as reproduction, offers a common pattern .... The variety of circumstances and the unity of function involved in reproduction offer a suggestive, loosely defined, and exceedingly complex range of formal comparisons."[14] Nor was the generative function selected at random: following Cesalpino, Linnaeus ascribed the "essence" of the plant to the parts of fructification.

It is not that Linnaeus refused to consider natural method. And at those times when he did so his conception of nature coincided with that expressed by Buffon: nature is infinitely nuanced, and our knowledge of it rests upon the number of distinctions any method can encompass. But while method thus offers the ideal reflection of nature, through a complete compendium of characteristics, Linnaeus points to the failure of every method so far to achieve this goal in practice. Meanwhile are we to refrain from our attempts at ordering nature? In addition, he voices doubts about the realm from which method largely derives its assorted resemblances and differences, this realm of *habitus*, of visible signs, the comparison of external appearances. It is just such groupings that remain unverifiable and unscientific. If a microscope is needed to make the "essential" identifications, then so be it. Comparisons made from *habitus* can only be considered secondary, while those made from the parts of fructification must be considered primary.[15] Where Linnaeus and Buffon part ways, then, is not over their idea of nature – its order, complexity or gradations – but over the limits of what we can know (the number of variables any method can handle), and the most adequate means of ordering nature (by one "essential" character or by a variety of comparisons according to size, shape and so on): their disagreement is over what constitutes a *science* of nature.

Buffon's application of the principle of continuity to nature, in his "Premier discours," was in fact nearly identical to Bonnet's, which he had already published in the "*Traité d'insectologie*," and would continue to elaborate in the *Contemplation*. Identical not only in application but also in its resultant logical complications; for the principle of continuity does not really achieve the consequences Buffon was after. As with Bonnet, this principle leads Buffon to admit to a nature too subtle and complex for our blunt mental capacities. We will never know it as it truly is, its gradations will never be our own. Yet he expresses nothing but disdain for systems because they begin, in effect, with this same premise: nature is too nuanced to be classified in the full wealth of all its relationships, hence, an artificial (but non-arbitrary) classification will have to do. Then as we have seen Buffon offers us a "natural" method for distinguishing species. Later still, in this same preliminary essay, amidst disparaging the arbitrariness of all systems, he again admits that nature cannot be known, and that it therefore matters little how we arrange things. From the human perspective, he suggests the most natural organization would be to start with man, moving on to domestic animals and to those most useful to man, then to those less useful, to those more exotic, and so on.[16] Buffon's debate, it would seem, is less with Linnaeus than himself. Can nature be known or not? Can beings be grouped together and ordered in a manner that truly represents their place in nature or not? When Buffon takes the offense against systems the answer is presumed to be yes, but when he returns to his thoughts on nature in all its concatenation the answer becomes no.

A still deeper logical problem – one raised neither by Buffon nor Bonnet – is this: if nature cannot be known, how can we presume to know even that it is unified, that it is a continuum of resemblances and differences? As a condition for its possibility, natural history depended upon the common affinity of things; whether the categories placed on nature were real or artificial, they depended upon the continuity of nature. For only continuity provided the assurance that nature was unified, that it was thus rational and orderable. This was the logical implication of Leibniz's "principle of sufficient reason". Were nature found to be discontinuous, it would also be found irrational.[17] To recall Bonnet's words: "If between any two beings there were an empty space, what reason would there be for passing from one to the other?" All matter, all beings, all worlds were arranged and ranked in this universal continuum, yet not in an attempt to prove the homogeneity of nature (for this was not questioned), but to reflect it, to represent it as nearly as possible in human terms and in human language.

The particular procedures of Bonnet emerge in his references to his own method. The naturalist, in his view, first gains access to some new facts which have been solidly confirmed; "he analyzes and atomizes them, compares them to one another and to facts already known", yielding to the consequences of his examination. "He then clearly exposes the order of these consequences; he binds them or rather they bind themselves." Do we wonder how it is that they so readily bind themselves? "Even in studying nature a little, one soon perceives that all its parts are closely linked through various affinities. It is the research of these affinities that must occupy anyone concerned with the natural sciences." The proper task of the naturalist is first to analyze the components and then to synthesize them. Bonnet insists that one step is as important as the other, and he is critical of the overly positivistic tendencies of many of his contemporaries who, he thinks, have misunderstood Newton, just as he is critical of the epigenesists who, he feels, speculate without any concern for observation. "To banish the art of conjecture entirely from the physical realm would reduce us to pure observation; and to what end would observation serve us, if we drew not the least consequence? We would ceaselessly amass material in order to not build. We would confound endlessly, the means with the ends. All would remain isolated in our minds whereas all is connected in the universe".[18] The practice of careful and regulated attention is only one aspect of observation: involved also is the "art of perceiving relations".[19] And those who grasp these general relations among beings will discover as well the harmony of the chain.

Reproduced below is Bonnet's first "idea" of the chain of beings:

MAN
Orangoutang
Ape
QUADRUPEDS
Flying squirrel
Bat
Ostrich
BIRDS
Aquatic birds
Amphibious birds
Flying fish
FISH
Crawling fish
Eels
Water snakes
Slugs
Snails
SHELLFISH

Tubular worms	Mushrooms, Agarics	Vitriols
Tinea	Truffles	METALS
INSECTS	Corals, Coraloids	SEMI-METALS
Gallflies	Lithophyte	SULPHURS
Tape-worm	Asbestos, Amianthus	Bitumens
Polyps	Talc, Gypsum, Selenites	EARTHS
Sea-anemone	Slates	Pure Earth
Sensitive	STONES	WATER
PLANTS	Formed Stones	AIR
Lichens	Crystallizations	FIRE
Molds	SALTS	More subtile matters[20]

In the *Contemplation*, where it is further elaborated, the scale underwent at least one modification: between plant and mineral, asbestos, for sharing some of the fibrous qualities of plants, has been moved next to the truffle, itself the lowest form of plant because it is a marbled and mottled mass, growing beneath the ground without roots, stem or leaves.[21]

The order, it should be noted, is chiefly a visual arrangement of similar appearances and exterior functions: the flying squirrel and the bat appear next to the ostrich, for example, because, while the first are the quadrupeds' contribution to the birds, the second is the birds' contribution to the quadrupeds. There is also a general principle operating in this distribution – that of "organization", in accordance with which Bonnet sets forth a quadripartite division of all beings: brute or unorganized; organized but inanimate; organized and animate; organized, animate and reasoning. "Of all the modifications of matter, the most excellent is *organization*. The most perfect organization is that which brings about the most effects with the fewest number of dissimilar parts." Thus emerges a hierarchy of corporeal perfections which would incorporate everything from the primary elements to man on the basis of structural complexity and differentiation of parts: "The being composed only of the repetition of resembling or similar parts or even in which one can conceive of distinct parts only by an operation of the mind, possesses the lowest degree of corporeal perfection. Such is probably the atom or elementary particle."[22] That distributions exist, which spread all things across a great and unified space, Bonnet never doubts. Yet he repeatedly draws back from claiming that his order of things is nature's own. "In sketching the chain of beings, I have not pretended to fix the gradations of nature: it is only a way of envisioning beings, of surveying them." And again: "But this very general view is only our way of conceiving and not the rule of things. We cast a glance only upon the exterior of beings, we traverse only their outer surface."[23] Shades of

Linnaeus' reservations about treating *habitus* as the conclusive taxonomic criterion.

Here a certain unspoken conflict arises, for what does one discern on that visible surface, constituting the discrete units of nature, but species – those of most enduring interest being the "transitional" species, which fill the breaks in the chain. This would present no difficulty were species to form only the basic units of our poor perception; but Bonnet, at times, also insists upon their actual and immutable existence, even associating such invariability with the highest sort of perfection. Yet the principle of continuity, taken to its logical end, finds in nature "as many steps as there are individuals": it results in nominalism; nature, as it truly is, lying in principle beyond our knowledge. And if we think in species rather than in individuals, that only serves to demonstrate the fact. Species invariability, on the other hand, while perhaps deriving from the same image of the universe as a designed creation, finds its support elsewhere, in theories of generation.

Such incompatibility of the continuity principle with that of species fixity has its most notorious example in Buffon. One can see why historians have been moved to comment, and perhaps to try to elevate Buffon above contradiction; for here we are confronted with two assumptions of equal importance to eighteenth-century natural history, and yet, when pressed, at odds with each other. This Buffon demonstrates, however inadvertently. Despite arguing, in his "Premier discours", that only individuals exist in nature, he nevertheless refers throughout it to species, assumes their existence, and, as we saw above, even presents a formula for species determination according to a tabulation of all similarities and differences. Such a definition does not, in fact, contradict the principle of continuity so long as these species remain no more than a tabular device. But Buffon did not stay with this definition long, taking up almost immediately, in the second volume (published in 1749, the same year as the "Premier discours"), what became his characteristic definition of species: species are real entities defined by the limits of reproduction as determined by the test of hybrid sterility. Already implied, and later made explicit, is that species not only are real, they are constant and invariable as well. (We will not dwell on the irony that hardly had Buffon finished taking Linnaeus to task for "arbitrarily" choosing reproduction as the basis for classification before he adopted it himself as the basis of species determination.)

Historians have not found it easy to accept that naturalists, such as Buffon or Bonnet, were unwilling to relinquish either species invariability or the continuity of nature. Lovejoy, for example, tries to have it that Buffon simply abandoned his preliminary effort to apply the principle of continuity to natural history: yet he goes on to acknowledge that there exist "many later passages where the old phraseology incongruously recurs", and eventually allows that Buffon was "trying to retain as much of the principle of continuity as could be made consistent with his present view".[24]

Bonnet's inconsistency is not so glaring, but that is only because his generation theory did not demand such a detailed examination of the species issue. Preformationism was premised, of course, upon the notion that organized beings had all been formed in a single act, at the creation of the world. Encased within the reproductive organs of all primordial females were the future generations of each species (*emboîtement*). The problem this theory solved – or skirted, depended upon how one saw the issue – was the difficulty of explaining the organizing principle of living beings. "When natural science has attempted to explain the formation of organized bodies *mechanically*, it has lost itself in the night of conjectures," Bonnet wrote, adding that one need not be Haller or Morgagni see connections between the parts of an animal so direct, complex and indissoluble that they must always have existed together.[25] The question of generation deferred, preformationists were free to concentrate on the mechanics of development, whereas the partisans of epigenesis were required to produce some organizing force that, it was hoped, could found its legitimacy on the prestigious model of Newton's gravitational force (Maupertuis, Buffon).

Supporters and opponents of preformation each went after the weakness in the other's position. Preformation was reproached for placing generation outside of science: rather than the looking for natural causes, it relied on God, and directed every supposed creation back to Creation. As Bonnet often said, there is no actual generation, only development. Wolff lamented such a stifling vision of things: "Hitherto it was a living nature, which through its own forces produced endless changes. Now it is a work which only appears to produce changes, but which in fact and in essence remains as unchanged as it was built, except that is gradually is more and more used up."[26] Well and good, except that the "forces" put forth as "natural" explanations of generation had no more empirical basis than did the reliance on God's creation. Nor was the logic sound. Wolff, for example, posited a *vis essentialis* of which he wrote, "It is enough for us to know that [this essential force] is there, and to recognize it from its effects, as it is demanded purely and simply in order to explain the development of parts. What we call it is still less important; only this must I remind you of, that it is that force through which in the vegetative body all those things are accomplished on the basis of which we attribute life to it."[27] To subtract the profundities and curlicues from *vis essentialis* is to be left with a sad tautology: living things live thanks to a life-constituting force, this force everywhere verified by the fact that living things are alive; and when they die, this force leaves them, as can be observed in their death and decomposition. As for Buffon's *moule intérieur*, he tried to associate it with the Newtonian model of universal gravitation: "There exists in Nature a power penetrating to the internal parts of bodies that we call gravity. I apprehend the idea of a *moule intérieur* as analogous to this power." But the functions of these two forces are very different: far from being a universal force, common to all matter, this interior mold was "a teleological force that arranged organic matter according to an intrinsic plan", and that influenced various organic particles differently.[28] As Malesherbes said of it, "to explain a fact, invent a word".

Hence the weakness of the preformationist explanation – in taking as a given what it could not explain – to a certain extent was also its strength. It did not have to trouble itself with elusive and contradictory forces that, despite their simplicity, arranged the complex structure of organized beings and faithfully preserved species traits. Paradoxically, the epigenesists came off looking like vitalists or mystics for clinging to the idea that a natural explanation of generation must be possible, while the preformationists ended up the mechanists.

But this was no standoff between empiricists (epigenesists) and dogmatists (preformationists), as the caricature once had it. So far as the facts went there really was little disagreement; however, these facts did not speak for themselves. Limited knowledge, rather than a failure to accept facts, certainly was one of the problems, and one which coincided, moreover, with the epistemological consideration that knowledge, in principle, had its limits. Here, quite possibly, was an area *essentially* unknowable. Was generation a subject that observation could ever resolve? – that was the question. Without cell theory providing a basic structural unit, there was no limit to the minuteness at which critical activities could take place. Looking for the secret through a microscope might only be chasing infinity. In addition, while organization was everywhere apparent in the animate being, its source was in no way clearly linked to any particular organ or element. It seemed to posit itself on the evidence of the structure as a whole. Like a wind that blows through the body, one witnessed it only in its effects. To try to read back from these effects to their source, to try to cage this force in a definition or hypothesis, led, as we saw, to rather thin results. Cell theory lent a point of focus and a basic structural unit from which to analyze the composition of beings. Henceforth, "Whatever the nature of an organism, whether an animal, a plant or a micro-organism, it is always made from the same elementary units. It is the arrangement, number and properties of the cells that give the organism its shape and qualities."[29]

This, it is said, finally ended the epigenesis-preformation debate. "The aspect of biology most profoundly transformed by the cell theory was the study of reproduction", according to Francois Jacob. "Gradually, improvement of the microscope and increasingly rigorous observation brought to a close the old argument about preformation and epigenesis .... What von Baer saw after fertilization of the egg was not the development of a little preformed organism, but a series of complex events from which the shapes and structures of the future adult gradually emerged."[30] That is the accepted view. As Stephen Jay Gould says, "We know that the epigeneticists had 'won' by von Baer's time", a view he appears reluctantly to accept, even as he goes on to undermine it:

> the solution to great arguments is usually close to the golden mean and this debate is no exception. Modern genetics is about as midway as it could be between the extreme formulations of the 18th century. The preformationists were right in asserting that some preexistence is the only refuge from mysticism. But they were mistaken in postulating preformed structure, for we have discovered coded instructions .... The epigeneticists, on the other hand, were correct in insisting that the visual appearance of development is no mere illusion.[31]

Along with this caution against finding a clear-cut victor, I believe one must also give up the notion that cell theory ended the argument, for it did not resolve the question of primordial formation. It did not clear up the problems of inheritance. Nearly a century later, for example, one finds Edmund Wilson, in his standard work, *The Cell in Development and Inheritance*, concluding with a section entitled, "Preformation and Epigenesis. The Unknown Factor in Development":

> We have now arrived at the furthest outposts of cell-research; and here we find ourselves confronted with the same unsolved problems before which the investigators of evolution have made a halt. For we must now inquire what is the guiding principle of embryological development that correlates its complex phenomena and directs them to a definite end. However we conceive the special mechanism of development, we cannot escape the conclusion that the power behind it is involved in the structure of the germ-plasm inherited from foregoing generations. What is the nature of this structure and how has it been acquired? To the first of these questions we have as yet no certain answer .... [It] raises once more the old puzzle of preformation or epigenesis. The pangen hypothesis of De Vires and Weismann recognizes the fact that development is epigenetic in its external features; but like Darwin's hypothesis of pangenesis, it is at bottom a theory of preformation, and Weismann expresses the conviction that an epigenetic development is

an impossibility. He thus explicitly adopts the view, long since suggested by Huxley, that "the process which in its superficial aspect is epigenesis appears in essence to be evolution in the modified sense adopted in Bonnet's later writings; and development is merely the expansion of a potential organism or 'original preformation' according to fixed laws."[32]

The problem, then, remained unresolved and much discussed at the end of the nineteenth century. In C. O. Whitman's well-known Woods Hole lectures on Bonnet (mid-nineties), the subject of the first lecture was not Bonnet so much as these same questions.[33] Why lecture on Bonnet anyway? Not out of historical interest, but to use his preformationism as the focal point for an elucidation of some essential distinctions in Whitman's own theory of development. If he is harder on Bonnet's system than would seem historically warranted, it is because Bonnet represents a still-challenging standpoint (or misconception) that must be addressed. We see from the Wilson quote what could even be called a swing back to the preformationist position, and the issue, as Whitman freely acknowledges, is that there still exists no established theory of heredity. Hypotheses abound on the "orientation" of the egg, whether it is definitely oriented from the beginning of its existence or not. This, Whitman declares, was the same question Bonnet asked – and it remains unanswered.[34] For his part, Whitman does not advocate a return to either eighteenth-century stance, Bonnet having claimed too much and Wolff too little for preformation. His position is, more or less, a preformationism divested of the creation hypothesis and the attendant *emboîtement ad infinitum*. "The indubitable fact on which we now build ... is the ready-formed, living germ, with an organization cut directly from a preexisting, parental organization of some kind." "Continuity of organization is indeed a species of preformation, but it is a preformation which no one will now deny .... To this extent we are all preformationists."[35]

Bonnet's preexisting germ calls to mind a miniature embryo with all the parts of the adult organism already formed and prepared to expand according to the mechanics of "development". In reality, and certainly this accounted for the persistent interest in his views, Bonnet tended, through the course of his writings, to modify this conception of the germ, abstracting it into an organizing principle. Yes, he supports *emboîtement*, but his version of it is refined until at times it hardly remains an hypothesis of encasement. From one of his later notes: "Germs are not enclosed one within another like boxes or cases: rather, a germ makes up part of another germ in the way that a seed makes up part of the plant on which it develops .... [T]he germs grow in and through one another."[36] Here only the thinnest thread required to link beings back to the original creation has been retained.

The nature of the germ itself is likewise generalized, from the visual image of a miniature individual to the logical requirement of a principle of organization. It is not really a subject for microscopes because the parts of the germ are so vastly altered in the course of development: "Were we able to see the germ [of a chick] enlarged, as it is when small, we would be unable to recognize it as a chick .... The parts of the germ do not all develop simultaneously and uniformly."[37] But these parts are not even visible. Only by degrees does the germ gell, from perfect transparency into the "living point" we first make out as the embryo; and since it is transparent, there literally is nothing to see. "Do not, therefore, mark the time when organized beings begin to exist from the time when they begin to become visible; and do not constrain nature by the strict limits of our senses and our instruments."[38] Bonnet came to employ general terms to describe an increasingly abstract concept: in the *Palingénésie*, the germ is thought of as a "representation" of the general organic whole; it is called a "primordial foundation"; it is now taken to designate "any sort of original preformation, from which an organic whole may result as its immediate principle".[39] Precisely the generality of this formulation kept it from aging; even with cell theory the nineteenth century was hardly in a position to make it more concrete.

Let us not claim perfect consistency for Bonnet in this view, for we must acknowledge that even in his later writings he sometimes fell back into a much more predelineated version of the germ. Still there is no doubt that he willingly detached much from the germ in order to preserve the notion of an inherited organization; and he opened the way for later writers to detach more, and perhaps (Whitman's complaint) to broaden the definition of preformation to the point of utter confusion.

For Bonnet, however, who would never have sacrificed the original creation, there was no means of conceiving a link between preformation and evolution. If the germ comes to represent the individual only in general, its ties to a fixed species are thereby strengthened: it is the basic species structure that remains constant. Already in the *Considérations*, though his notion of *emboîtement* is more strictly drawn here, Bonnet connects the germ's organizing structure to species traits rather than to individual characteristics:

> The germ carries the original imprint of the species, and not that of the individual. It is a man, a horse, a bull, in miniature, but it is not a *certain* man, a *certain* horse, a *certain* bull. All the germs are contemporaneous in the system of *emboîtement*. The characteristics are not transmitted from one to another. I do not say all members are exactly similar. I see nothing of identity in Nature, and without having recourse to the principle of indiscernibles it is quite clear that the germs of the same species do not all

develop in the same uterus, at the same time, in the same place, in the same climate; in a word – in the same circumstances. Here are the causes of individual variation. The most efficacious single cause is the seminal liquor.[40]

Variations due to environmental differences, nutrition, the male's seminal fluid – which acts both as a powerful stimulant capable of inducing the female egg to grow, and as the bearer of male characteristics capable of marking the offspring – such factors account for individual characteristics, for monstrosities and even for some remarkable variations within the species. But these concessions in no way effect species stability. "Nature is surely admirable in her conservation of individuals, but she is especially so in the conservation of species .... Nothing changes, nothing alters; perfect identity. Victorious over the elements, time and death, the species are conserved ...."[41] And to this perception Bonnet remained faithful. Ten years after completing the *Palingénésie*, he wrote, "I have never attributed the *formation* of the least thing to *accident*. I have always conceded and maintained the preformation of everything that is truly *organic* .... In a word – *and can I repeat it too often*? – I have never conceded anything but simple modification of preformed parts, except certain cases of grafts or accidental separations."[42] Species are thus preserved, by means of preexisting germs, even in the face of monsters and mutants.

Bonnet's fundamental dimension is that of space. Time has nothing intrinsically to do with the gradation of beings, it has no fundamental role in the system. When he makes use of time, it is only as another factor of classification; it is the field upon which the creation unfolds. There is no evolution. As Foucault has put it: "time is never conceived as a principle of development for living beings in their internal organization; it is perceived only as the possible bearer of a revolution in the external space in which they live".[43] Catastrophes, world revolutions (the last being at the time of the Mosaic flood) are predetermined as much as is any preexisting germ. Moreover, under Bonnet's guidance all changes through time become evidences of progress which in turn are made to conform to the law of continuity, so that the change of any one species displaces the entire scale – and the only reasonable shift is toward the perfection of God. Whether we look to the catastrophic world revolutions (of which, in Bonnet's estimation, the world had already been made to undergo at least two) or to those passages in which gradual change is considered, the direction of the movement, by prearranged plan, is invariably toward perfection:

there will be a continual and more or less slow progress of all the species toward a higher perfection, so that all the degrees of the scale will be continually variable within a fixed and constant relation. I mean to say that the mutability of each degree will always have its reason in the degree immediately preceding it . . . . Man, once transported to an abode better suited to the eminence of his faculties, will leave to the ape or the elephant that preeminent place which he formerly occupied among the animals of our planet. In this universal restoration of animals, therefore, it will be possible to find Leibnitzes and Newtons among the apes or the elephants, Perraults and Vaubans among the beavers, and so on. In this new hierarchy, inferior species, such as oysters and polyps, will stand in the same relation to higher species as birds and quadrupeds do to man in the present hierarchy.[44]

Time has been employed here only to complete the continuity of space and to turn the chain of beings into a slow-moving escalator, rising toward that one Being who stands outside of it. For Bonnet, it was only another tribute to the perfection of the entire concept that there were an infinite number of gradations between God and the highest species on the scale so that this process of enforced perfection could continue virtually forever.[45]

The moment has come to recall the ever useful orang from the wings and parade him once more about the stage, noting this time, however – despite Bonnet's protestations of similitude – a certain ungainly dullness, his bizarre and, in the end, only caricatural resemblance to the ordinary *valet-de-chambre*: let us mark for all time then that gulf, never to be bridged, separating man and ape. For where is his Reason?

But why do we not hear Bonnet protest this point after all? Why is it in fact his own distinction? He ought to, it would seem, as Robinet complained, for how does the great eighteenth-century proponent of the principle of continuity hope to maintain his continuum if, at the same time he introduces an attribute applying absolutely at one level of the scale, it is found to be totally nonexistent at the very next level? "The negative is always an infinite distance from the positive", Robinet concluded in a sweeping logical stroke, and not, as might be expected, in order to reduce the principle of continuity to an absurdity, but to save it: "For myself I would rather give even intelligence to the least atom of matter – provided it were in a degree and of a quality suitable to it – than refuse organization to the fossils and make of them isolated beings, having no connection with others. It is to no purpose to tell me that this is a bizarre opinion, and that it is not possible that a stone thinks . . . . [I]f the law of continuity is admitted, we ought likewise to admit all that follows from it; while it is inexcusable to abandon so general a principle without a sufficient reason."[46]

And this "infinite distance" separating man from ape is only a beginning. As for the rest of the quadripartite division of nature accepted by Bonnet, what remains after the force of Robinet's criticism has been felt? Are they not each differences in kind, not in degree? And are we not led, of necessity, therefore, by following this logic, to discontinuity?

> What continuity can there be between the organic and the inorganic, between the animate and the inanimate, between the rational and the non-rational? It is evident that there is no mean between the positive and the negative, and consequently, that there are no intermediate beings which link the two together. If there were such beings, it would be necessary that their constitution should simultaneously participate in two mutually exclusive contraries; . . . e.g., that the passage from inorganic to organic should be filled up by a middle sort of beings which are both organic and inorganic. But such beings are self-contradictory (*se repugnent*). If we wish to leave the law of continuity standing, . . . if we wish to allow Nature to pass insensibly from one of her productions to another, without compelling her to make leaps, we must not admit the existence of any inorganic beings, or an inanimate, or any nonrational.[47]

If Robinet's answer was to extend all qualitative distinctions until they spanned the entire grid of the potentially classifiable, attenuating gradually in one direction or another, Bonnet seems to have remained rather oblivious to the whole problem; and perhaps the blame or credit should go to Leibniz, whose "project of establishing a mathematics of qualitative orders is situated at the very heart of Classical thought; its gravitational center".[48] In this sense, perhaps Robinet was more out of step with his times than Bonnet; yet the future has been foreshadowed. The final rupture came not at the upper end of the scale between man and monkey but at its base. It focused on the one particular link of the chain that had always been the weakest anyway, and drove a wedge so deep that the resulting breach split forever the continuity of nature into a radical dichotomy of organic and inorganic.

In spite of his rejection of the chain of beings and, indeed, his condemnation of all philosophical speculation on nature, Georges Cuvier (1769–1832) still felt the enchantment of Bonnet's system. "This idea", he repeated whenever speaking of Bonnet, "has grandeur and seizes the imagination". But a closer examination revealed its weaknesses, and in particular Cuvier looked to the first large division at the base of the structure rather than to the one between the uppermost species:

> Bonnet put at the bottom of the chain of beings the substances not yet organized. He considered crystal as beginning to have some organization . . . . He imagined, for example, that between asbestos and the mushroom or another of the last plants on the chain there was only a slight distance. However, when one carefully examines the nature of these kinds of beings, when one sees that asbestos presents only silken fibers resembling, visually, tendons or other parts of the animal body . . . when one sees that it was tested

neither for expansion nor diminution, *that it is not susceptible to death because it is not susceptible to life*, and when one sees to the contrary, that the simplest plants, those most similar in appearance to non-organized matter, like the truffle and the lowest mushroom, are born always from a seed; that during their whole existence they are increasing or diminishing; that they subsist only by absorbing heterogeneous substances, which act and react one on the other; that after a certain time they are destroyed by a spontaneous death; finally that after this death they dissolve, not into clay or earth like asbestos and amianthus but into aeriform substances or liquids, one is left convinced that between asbestos or amianthus and the simplest plants there exists only an apparent resemblance .... There is only a visual resemblance. The composition, the manner of birth, the manner of subsisting and perishing, are absolutely different. The idea of a passage from mineral to plant is superficial and false.[49]

Two reactions present themselves immediately upon reading this passage: first, that this certainly is a just criticism; yet, secondly, one cannot help but be persuaded that Bonnet did not need it pointed out, that he knew all of this very well already. The man who had all but blinded himself by the age of twenty-five in his attempt to come to grips with the processes of respiration in plants, of nutrition and especially of generation in animals – that is to say, the processes of life – who spent years poring over the complexity of organized beings, in their forms but also in the interrelated and interdependent nature of their functions: who knew better the fundamental distinctness of living from inorganic matter? Indeed it was on these very grounds, and not on some Christian stance, that he disregarded rather than embraced the work of the epigenesists, who would have run a modicum of life (or rather its essence) down into the lowest rungs of his chain, expressing it as forces of matter and motion, who would have thus preserved rather than disturbed the continuity of his chain, cinching the connection he allegedly sought.[50] But he would have none of it because no mechanical force seemed capable of forming such organized structures as life presented.[51] That is why he maintained his argument of preexisting germs, which again suggested from the beginning a much more radical dichotomy between organic and inorganic matter than he ever really faced up to.[52] One wonders, for instance, how Bonnet could so easily pass over this difficulty, especially when his own theory of preexisting germs cements it: "It is necessary to admit that this is not so satisfactory a transition as those observable among several other classes of terrestial beings: Nature seems to make a leap here; but this leap will disappear, without doubt, when our understanding acquires greater scope and precision."[53]

Do we not find Bonnet caught in a contradiction at this point? On the one hand, there is the metaphysical assumption that makes feasible his system of ordering and establishing a continuum in the first place – that nature is fundamentally a unity, that all similarities binding and dissimilarities distinguishing are arranged on a homogeneous background of resemblances. On the other hand, he refuses to let slip away the one thing his experiments have taught him to be true: that living matter, in its ability to form organized beings with such regularity, with such delicate precision, with such complexity as to the development of a single organ, not to mention the interrelated complexity of an entire system so obviously constructed, not piecemeal or haphazardly, but according to one unifying play; that, in short, his experiments have taught him living beings are something quite different from inorganic matter. Cuvier reprimanded Bonnet on this point, but the chain had already begun to crack beneath the tension found within Bonnet himself. The interstice between mineral and vegetable, pointed to casually by Bonnet as he unfolds his quadripartite system in the *Contemplation*, has by the end of his last works begun to take on the appearance of an open rupture, rendering the old four-part division meaningless and substituting instead a more radical division of the living and the nonliving.

Thus we find the dichotomy more persistently present in the *Tableau*. And it appears there in the same place that it will emerge and become a bold pronouncement in the works of Cuvier – interwoven into the fabric of a discussion on the complex and intimate structural relationship of organs, a first approximation of that "essential truth being sought, the irreducible structural and active elements of vital organization".[54]

Now I entreat you to tell me what mechanism will preside in the formation of a brain, a heart, a lung, and so many other organs? Yet I have not expressed the most salient difficulty: it consists not only in the mechanical formation of this or that organ, itself composed of so many different parts; it consists principally of producing, through the laws of mechanism alone, this throng of varied relationships that so tightly bind all the organic parts, and by virtue of which they all conspire to the same general goal; to form this unity called an *animal*, this organized whole that lives, grows, senses, moves, maintains itself, and reproduces.

Take note that the brain implies the heart, and the heart, in its turn, the brain. The brain and heart imply the nerves, arteries and veins. But the animal feeds; the organs of circulation thus imply those of nutrition. But the animal moves; the organs of movement assume therefore those of sensation. But the animal propagates itself; the organs of generation imply again those of nutrition, circulation, sensation, and movement. Clinging to generalities must be avoided here; detail, the greatest of detail, must be gone into.[55]

Although this is not Cuvier, although the functions of the organs have not yet been made more important than the organs themselves,[56] nor as yet ranked in a hierarchy of levels and dependences, the fact of coexistence and an interdependence according to a single unified plan has become a central aspect of defining the organism. There is recognition of the need for more than simply an understanding of the visible surface arrangements of an animal in order to establish its identity, that one must penetrate to the interior regions in order to appreciate and identify the true nature of a living organism. Bonnet says of the epigenesists: if they would do no more than simply look closely at a caterpillar, "they would see, not without a profound astonishment, some four thousand muscles employed in the construction of the caterpillar, their admirable coordination, those trachea no less admirable, and I like to persuade myself that they would then sense that a totality so prodigiously composed and yet so harmonious, so essentially one, could not have been formed, like a watch, from corresponding pieces, or from the meshing of an infinity of diverse molecules united by successive apposition."[57]

Professor Lovejoy has written of Bonnet in connection with the temporalization of the chain of beings. Yet time, at this juncture, does not seem to supply the important key to unlocking the chain or to breaking it in preparation for true evolutionary thought. Bonnet demonstrated, and others did as well, that one could have time but not evolution; one could have mutation too and not evolution. Bonnet's importance may perhaps be better measured by the extent to which he was able to bring the chain of beings to life; never mind if this was not entirely intentional on his part or if it wrought irreparable damage to his continuum.

In the *Palingénésie* he seems to guarantee the sharp dividing line between living and nonliving things by shifting plants further to the side of the living. Under the categories of the quadripartite division, as Professor Savioz has summarized it, "An organized being lives a *vegetative* life if its vital functions amount to nutrition, growth, and reproduction: the life of a *plant*. It possesses a *vegetative* and *sensitive* life if it has the *sensation* of its actions: the life of an *animal*."[58] Sensitivity broke off somewhere around the mussel. But in the *Palingénésie* there is a new emphasis:

> In general, one has great difficulty being persuaded that plants could be sentient beings. Since they never move about and since their forms have nothing in common with those of animals best known to us, there are no grounds for believing that they could participate a little in animality; grounds, in effect, for the conjecture that some relation exists between a violet and a butterfly, between a pear and a horse!
>
> We ordinarily judge beings only by such coarse comparisons. We compare them rough and wholesale in their form and in their structure, and if this superficial examination offers us no trace of similitude, we find ourselves at a loss to imagine one . . . .
>
> But as the locomotive faculty accounts for much in the perfection of organized and

sentient beings, were the plant endowed with some sensitivity, were it a perfectible being, there is reason to think that in its new state it will be capable of transporting itself from one place to another in accordance with its desires.[59]

In this passage the combination of themes we have been observing come together and are interwoven in a way that makes clear the paradox of how the study of comparative anatomy, on the one hand, could have been encouraged by adherents of the chain of beings in order to fill the hiatuses while, on the other hand, it served to bring about its destruction. The plant now participates in a certain "animality", and in the space of that admission the truffle moves an infinite distance from asbestos; not simply because, as it may at first appear, the vegetable moves into a closer proximity to the animal – for there are qualitative distinctions up and down the chain – but because it has been brought into the fold of the living; and even that shift is important only to the extent that what is classifiable is itself in the process of shifting. Taxonomy more and more clearly applies to living beings alone, and to them only because they are alive. If the reason for ordering pears in relation to horses holds some greater degree of justification, the rationale for classifying rocks with violets disappears.

Bonnet has begun to move out of the sphere of natural history into that of biology. Here I use Foucault's emphasis and rely on the importance he puts on this shift. "Historians," he claims, "want to write histories of biology in the eighteenth century; but they do not realize that biology did not exist then .... And that, if biology was unknown, there was a very simple reason for it: that life itself did not exist. All that existed was living beings".[60] Life for the natural historian did not constitute an obvious threshold. It slid up and down the scale: with Maupertuis, for example, defining life by mobility and relations of affinity, life was found in the most simple particles, whereas Linnaeus set it much higher. The question of the nature of life was not an important one, for this was not the naturalist's concern. His concern was with classifying, and life entered only as another category in this process, remaining relative to the chosen criteria as did every other category.[61]

As for the opposition that develops between the organic and the inorganic, it is not to be understood as the triumph of vitalism over mechanism. Rather, the methods of taxonomy have been modified; organic structures come to be seen as discontinuous entities; relations cease to be drawn in the visible world of identities and differences, but come to depend instead upon internal unities where classification is determined by functions performed by the organs. The closer we approach the more general classes, Cuvier points out, "the more the properties that remain common are constant; and as the most constant relations are those that pertain to the most important parts, so the characters of the higher divisions will be found to be drawn from the most important parts".[62] When the order of classification is derived from the essential functions of the primary organs, it becomes understandable how the very possibility of classifying is made dependent upon the fact that what is being classified is alive.

> Life is no longer that which can be distinguished in a more or less certain fashion from the mechanical; it is that in which all the possible distinctions between living beings have their basis. It is this transition from the taxonomic to the synthetic notion of life which is indicated, in the chronology of ideas and sciences, by the recrudescence, in the early nineteenth century, of vitalistic themes . . . what is being established at this particular moment is the conditions of possibility of a *biology*.[63]

By the very fact that they are alive, beings can no longer be ranged in graduated scales but "must group themselves around nuclei of coherence which are totally distinct from one another."[64] Continuity is reestablished by comparative study of the functions of the greater and lesser organs, yet "Nineteenth-century nature is discontinuous exactly in so far as it is alive."[65]

We have moved, then, beyond the conceptual structure of natural history, in which the mineral and the rock were as easily ordered as living beings, in which all was ranged on the same scale, in a shared space, on to a mode of thinking that made biology, the science of life, possible. Where Bonnet fits into the one, and where he seems to anticipate the other, I have tried to give some indication.

NOTES

[1] *Contemplation*, pt. III, Chapter XXX, 175–6n. The *Contemplation* was first published in 1764, yet much of it, including part III, had originally been written 1747–52, so that about thirty years separate the text from the notes, composed 1779–81 for the collected edition.

[2] *Ibid*.

[3] *Ibid*., part IV, Chapter III, 184n.

[4] *Ibid*., part II, Chapter IX, 51.

[5] *Ibid*.

[6] *Ibid*., part III, Chapter XXX, 175.

[7] *Ibid*., part II, Chapter XI, 53.

[8] Ernst Cassirer, *The Philosophy of the Enlightenment* (Boston, 1955), 61–4.

[9] Foucault, *The Order of Things*, 268.

[10] *Ibid*., 162.

[11] *Ibid.*, 137.
[12] *Ibid.*, 144.
[13] John Lyon, "The 'Initial Discourse' to Buffon's *Histoire naturelle*: The First Complete English Translation", *Journal of the History of Biology*, **9**: 1 (Spring, 1976), 148–52, 154–55.
[14] James L. Larson, "Linnaeus and the Natural Method", *Isis*, **58**: 193 (Fall, 1967), 305.
[15] *Ibid.*, 313–14.
[16] Buffon prefaces this classificatory scheme, which he calls "the most natural of all", by saying: "With regard to the general order and the method of distribution of the various subjects of natural history, this could be considered purely arbitrary. Consequently, one is certainly free to choose what seems either the most convenient or the most commonly accepted." "Initial Discourse", 161–2.
[17] Arthur O. Lovejoy, "Buffon and the Problem of Science", *Forerunners of Darwin, 1745–1859*, Bentley Glass, Owsei Temkin, and William L. Straus, Jr. (eds.), (Baltimore, 1959), 89; Robert A. Nisbet, *Social Change and History* (New York, 1969), 174–8.
[18] *Tableau*, XV, 67–70, 73–4, 75.
[19] *Considérations*, V, Chapter IX, art. CLIII, 260.
[20] *Oeuvres*, I, facing p. 1.
[21] *Contemplation*, part III, Chapter VI.
[22] *Ibid.*, part II, Chapters I and III.
[23] *Ibid.*, Chapter XIII.
[24] "Buffon and the Problem of Species", 92–3, 96, 100. For more on Buffon, I recommend Roger's *Les Sciences de la vie*, 527–84; Paul L. Farber, "Buffon and the Concept of Species," *Journal of the History of Biology*, **5**: 2 (Fall, 1972); Philip R. Sloan, "The Buffon-Linnaeus Controversy, *Isis*, **67**: 238 (September, 1976); Lyon's introduction to his translation of the "Premier Discourse"; Peter J. Bowler, "Bonnet and Buffon: Theories of Generation and the Problem of Species", *Journal of the History of Biology*, **6**: 2 (Fall, 1973); and Robert Wohl, "Buffon and his Project for a New Science", *Isis*, **51**: 2 (June, 1960).
[25] *Contemplation*, part VII, Chapter VIII.
[26] From Wolff's *Theorie von der Generation in zwo Abhandlungen erklart und bewiesen* (Berlin, 1764), 73, quoted in Roe, "Rationalism and Embryology," 17.
[27] *Theorie von der Generation*, 160; quoted in "Rationalism and Embryology", **23**.
[28] Farber, "Buffon and the Concept of Species", 263–4.
[29] Francois Jacob, *The Logic of Life: A History of Heredity* (New York, 1976), 120–1. Bonnet specifically doubts, allowing for every possible perfection of the microscope, that we will one day see the most simple substances of which everything is made. *Palingénésie*, part XIII, Chapter II; part XII, Chapter VII. Parts XII and XIII, in fact, comprise an inventory of the current imperfections and the enduring limits of our knowledge, a topic that haunts Bonnet. Here the list of our essential ignorances is so great that, to Bonnet, it indicates once again our still-lowly place in the chain: he concludes that, since we understand so little of it, the world was not made for us but for superior intelligences.
[30] *The Logic of Life*, 121.
[31] Stephen Jay Gould, *Ontogeny and Phylogeny* (Harvard, 1977), 18.
[32] Edmund B. Wilson, *The Cell in Development and Inheritance* (New York, 1898), 327

–8. The quote of T. H. Huxley's is from *Science and Culture* (New York, 1882), 295–6. Later editions of Wilson's book grew mightily, but the problem remains as unresolved as ever into the first decades of this century. Thus in the third edition of *The Cell* (New York, 1928), the book starts off with a discussion of this problem, mentioning Bonnet in the "Introduction" as well as C. O. Whitman's lecture, which we will take up directly; and it ends with the problem, retaining the section on preformation and epigenesis, which now begins:

We have now arrived at the furthest outposts of investigation in this field and may more clearly perceive the modern aspect of the old controversies that grew up about the problem of preformation and epigenesis. Perhaps this problem is scientifically insoluble; at any rate no general agreement has yet been reached in regard to it. The modern biologist must be permitted to treat the problem in pragmatic fashion, employing the terms of one hypothesis or the other according to the procedure that he finds most useful in practice. The physiologist and experimentalist have for the most part instinctively turned towards an epigenetic interpretation of development, the morphologist towards a preformistic or evolutionary one .... This does not, of course, mean the acceptance of preformation in the older and cruder sense. It means (to paraphrase the expression of Driesch) that we are unable to conceive how a self-determining system can increase its own initial complexity by interaction of its chemical and physical components.

Huxley's passage is retained for the reason that "we cannot fail to perceive that in certain respects the conclusion expressed by Huxley approaches the results of modern experimental research". (1110)

33 C. O. Whitman, "Evolution and Epigenesis", *Biological Lectures, Woods Hole* (1894).

34 *Ibid*., 209–10.

35 *Ibid*., 213.

36 *Contemplation*, part VII, Chapter IX, note 2. This is the view of Bonnet also taken by Wilson, who writes in the "Introduction" to the third edition of *The Cell*: "It was the service of Bonnet to push this conception to its logical consequence, the theory of *emboîtement* or encasement, and thus to demonstrate the absurdity of its grosser forms; for if the egg contains a complete embryo, this must itself contain eggs for the next generation, these other eggs in their turn, and so *ad infinitum*. ... Bonnet himself renounced this doctrine in his later writings, and Caspar Friedrich Wolff (1759) led the way in a return to the teachings of Harvey, showing by precise actual observation that the egg does not at first contain any formed embryo whatever; that the structure is wholly different from that of the adult ..."[6]

37 *Considérations, Oeuvres*, V, Chapter VII and IX.

38 *Ibid*., art. CLXXII.

39 Part X. Chapter I; part XI, Chapter III; part X, Chapter II. Doubtless it was this last passage regarding "any sort of original preformation" that Huxley accepted as Bonnet's later position. Bentley Glass, in his "Heredity and Variation in the Eighteenth Century", *Forerunners of Darwin*, 167, passes off Huxley's attribution to Bonnet of the position that there need be no actual miniature organisms but only an "original preformation" as only Huxley's "rendering" of Bonnet. Glass adds: "The actual statement reads: 'I add here that I understand in general by the word germ every preordination, every preformation of parts capable by itself of determining the existence of a Plant or of an Animal.'" Were this the statement, then "rendering" would perhaps be the correct term, but

obviously it is not. The view of Huxley (and of Wilson following him), I would have to conclude, is really more faithful to Bonnet than that of Whitman (and of Glass following him), which would ignore or suppress the much broader interpretation of the germ found in Bonnet's later writings. To this it would have to be added, however, that Bonnet was not exactly consistent, and that more rigid analyses often alternated with those from which a "later" Bonnet emerges.

40 *Considérations, Oeuvres*, VI, art. LXV, 392–3.

41 *Ibid.*, V, art. CXL, 231–2.

42 *Ibid.*, VI, art. CCCLX, note 1.

43 Foucault, *The Order of Things*, 150.

44 *Palingénésie*, XV, part III, Chapter III, 218–20.

45 It was this aspect of Bonnet's thought that inspired French traditionalists like Louis de Bonald, Joseph de Maistre and Pierre Ballanche to turn to him in the first half of the nineteenth century for a theory of progress and social change that at the same time offered an alternative to the Revolution and all it implied. Ballanche borrowed the term *palingénésie*, which meant "new birth" or "natural evolution" as Bonnet used it, and wrote *Essais de palingénésie social*; see Michael Reardon, "Pierre Ballanche as a French Traditionalist", *Catholic Historical Review*, 53 (January, 1968), 573–99.

46 Quoted in Arthur O. Lovejoy, *The Great Chain of Being* ([1936], New York, 1960), 277.

47 Quoted *ibid.*, 276.

48 Foucault, *The Order of Things*, 57.

49 Georges Cuvier, *Histoire des sciences naturelles* (Paris, 1843), IV, 251–2 (my emphasis).

50 Bonnet's approach is certainly not that of a religious dogmatist: He defends himself on scientific rather than on theological grounds. This, Cuvier, a no-nonsense observer of nature himself, points out with admiration: "what is remarkable in Bonnet . . . is that neither religion nor ethics are attained there as they are in La Mettrie and others" (*ibid.*, 261).

51 "All force is indeterminate and cannot of itself produce specific effects, or in other words it presupposes the existence of the mechanism which it activates. If, therefore, there is nothing preformed in the matter which the essential force is supposed to organize, how can that force produce, let us say, an animal instead of a plant, a particular kind of animal, and a special organ appropriate to that animal?" (*Considérations*, part II, Chapter VII, 437–8n).

52 Even if one takes into consideration the potential or gradual advances of nature from one revolution to the next, this process involves only germs already preformed. A rock, for example, which has no preexisting germs, can never advance.

53 *Contemplation*, part III, Chapter V.

54 William Coleman, *Biology in the Nineteenth Century* (New York, 1971), 21.

55 *Tableau*, 95–6.

56 Cuvier says we must focus "rather upon the functions than upon the organs". Quoted in Foucault, *The Order of Things*, 264.

57 *Tableau*, 97.

58 Savioz, *La Philosophie de Charles Bonnet de Genève*, 94; also above, page 40.

59 *Palingénésie*, 236, 242.

60 Foucault, *The Order of Things*, 127–8.

[61] *Ibid*., 160–1.
[62] Quoted *ibid*., 228.
[63] *Ibid*., 269.
[64] *Ibid*., 272.
[65] *Ibid*., 273.

CHAPTER III

SIGNS AND IDEAS

Eighteenth-century considerations of language share a certain concept of the linguistic sign and its representative role in signification as well as a concept of the way in which this representative capacity of signs establishes speech and thought in their theoretical possibility and in their actual genesis. Yet this fact, fundamental though it may be, requires a certain amount of careful attention. For if the sign system is dominant its presence is also unobtrusive.[1] Little consideration was given, for instance, to definitions of what was intended by a sign, something apparently not felt to be required as a point of departure when employing sign theory.[2] D'Alembert's article "Signe" in the *Encyclopédie*, which *does* provide a definition, simply pieces together two passages, one from the *Logique de Port-Royal* and the other from Condillac's *Essai sur l'origine des connaissances*:

> The sign is everything destined to represent a thing. The sign encloses two ideas, the one of the thing that represents, the other of the thing represented; and its nature consists in exciting the second by means of the first. Various divisions of signs can be made, but we are content here with three that are of the greatest utility. I distinguish three kinds of signs: (1), accidental signs, where particular circumstances have linked objects to our ideas in such a way that they are suitable to revive them; (2), natural signs, or the cries that nature has established for the sentiments of joy, fear, pain, etc.; (3), institutional signs, or those we have chosen ourselves, and which have only an arbitrary relation to our ideas. These last signs are necessary to man in order that the exercise of his imagination be in his power.

No treatises were written with the exclusive intention of presenting a general theory of the sign. Instead the sign will be encountered in studies on logic and on the operations of the understanding, its properties may be analyzed in a treatise on etymology, and it will again appear in the innumerable discussions of the origins of language and in the prefaces of grammars. It is not singled out.[3]

On the other hand, d'Alembert's brief definition was perhaps sufficient; or, more to the point, the two sources from which he drew his definition were not chosen carelessly or at random, for these accounts of the function of signs directed the discussion of sign theory in the latter half of the eighteenth century.

In the fifth edition (1683) of the *Logique de Port-Royal*, two chapters were added that dealt with the relation between the sign and the signified. The principal discussion was inserted into the first part as chapter four: introduced, that is to say, in the midst of a consideration on ideas. And in what did this consideration consist? The preeminent characteristic of ideas, we are informed from the start, is their representative capacity. Part I is entitled "Conception", and from the preceding Introduction we learn that "To conceive a thing is simply to view that thing as it presents itself to the mind. For example, we are conceiving when we represent to ourselves a sun, an earth, a tree .... The form by which we represent a thing to ourselves is called an idea."[4] In the course of Part I ideas are considered in five ways: according to their nature and origin, according to the kinds of objects they represent, according to their simplicity or compositeness, according to their extension (their generality, particularity, or singularity), and according to their clarity or obscurity. Not by accident is the chapter on signs inserted following the classification of ideas according to their objects, where the relation of the representation to the object is at issue. The chapter opens:

> When one considers an object in itself and according to its own being, while ignoring what the object is able to represent, the idea he has of it is an idea of thing, such as the idea of the earth or the sun. But when one considers a certain object only as representing another, the idea he has of it is the idea of sign, and the first object is called sign. Maps and pictures are ordinarily regarded as signs. Thus the sign encloses two ideas, the one of the thing that represents, the other of the thing represented; and its nature consists in exciting the second by means of the first.[5]

What we witness with the sign, then, is a duplicated representation, the overlaying of a representative idea with a representative sign. Although the sign always in some way remains distinct from what it signifies, by its very nature the idea of the thing signifying excites the idea of the thing signified; and the sign subsists so long as this double idea is excited.[6]

Now the sign chosen by the Port-Royal logicians for their first example is not a word or mark or mathematical symbol, but a map or picture, a choice so well suited as an illustration of representation that, to a representative view of language, it proved a rich metaphor, often repeated during the next hundred years, of language as a painting or copy of the mind. In the opening passage of the article "Grammaire" in the *Encyclopédie* we read, for example, that "Speech is a sort of picture or copy (*tableau*) of which the original is thought; it must faithfully imitate it, insofar as fidelity can be found in the sensible representation [i.e., sound-sign] of something purely spiritual."[7]

The metaphor's aptness finds a complementary reason in the rational principles guiding not only the *Logique* but equally the *Grammaire de Port-Royal*. The title given to the *Logique* is the "art of thinking": "art" because it is always the task of art to dispense rules, and "art of thinking" rather than the more traditional designation "art of thinking well" because there is no art of thinking badly. Rules determine correct procedure; incorrect thinking is unruled thought. This principle of incorrect thought as non-thought extends to the *Grammaire*, which is now simply the "art of speaking", for effective speaking exists only according to rules. But what rules, or whose? Here an important consequence follows: "grammar will not be valued like the prescriptions of a legislator finally giving to the disorder of speech its constitution and laws .... It is a discipline that enunciates the rules by which a language is of necessity regulated for it to be able to exist. It has to define that correctness of a language which is neither its ideal, nor its better usage, nor the limit that good taste cannot cross, but the form and internal law which very simply allows it to be the language it is".[8] Grammar, in consequence of the Port-Royal treatment, takes on a double meaning; it must uncover rules that do not share the same level: "There is a grammar which is the immanent order of all pronounced speech and a grammar which is the description, analysis and explication – the theory – of this order."[9] From which devolves the bifurcation of grammar into its universal and particular pursuits:

> Grammar thus admits two types of principles. The first are of an immutable truth and of a universal usage; they hold to the nature of thought itself; they follow upon and are only the result of the analysis of thought. The rest have only hypothetical truth and derive from free and mutable conventions, and their use derives only from the peoples that have freely adopted it, without losing the right to abandon or change them .... The first constitute *grammaire générale*, the others are the object of various *grammaires particulières*. *Grammaire générale* is thus the reasoned science of immutable and general principles of pronounced or written speech in every language. A *grammaire particulière* is the art of applying the arbitrary and customary institutions of a particular language to the immutable and general principles of pronounced or written speech.[10]

There exist rules for thought, and only thinking correctly is thinking at all. In the same way, discourse has its level of universal rules, coincident with those of thought, assuring for signs the representative accuracy that a geographic map bears in relation to a country with its rivers, cities and boundaries; and it is these rules which the science of *grammaire générale* is assigned to uncover.

The notion of the sign is fully operative in the *Grammaire*, but no theory of it appears there. That comes, as we know, in the *Logique*. Foucault considers the discussion of signs found there sufficient, by itself, to account for the new sign system constitutive of Classical thought, and for this reason:

> Ever since the Stoics, the system of signs in the Western world has been a ternary one, for it was recognized as containing the significant, the signified, and the 'conjuncture' (the τύγχανον). From the seventeenth century, on the other hand, the arrangement of signs was to become binary, since it was defined, with Port-Royal, as the connection of a significant and a signified .... The relation of the sign to the signified now resides in a space in which there is no longer any intermediary figure to connect them: what connects them is a bond established, inside knowledge, between the *idea of one thing* and the *idea of another*.[11]

The new arrangement presented a new problem – and new solutions; "in the sixteenth century, one asked oneself how it was possible to know that a sign did in fact designate what it signified; from the seventeenth century, one began to ask how a sign could be linked to what is signified. A question to which the Classical period was to reply by the analysis of representation; and to which modern thought was to reply by the analysis of meaning and signification".[12]

There was, however, an aspect of sign theory central to eighteenth-century considerations with regard to which the *Logique*, although it provided some important reflections, was nevertheless unsatisfactory and had to be revised to accord with the impact of Lockean psychology. And it is here that the reason for d'Alembert's grafting of the *Logique* to the *Essai sur l'origine des connaissances* can be found. What we are referring to is the last of three "very useful divisions of signs" that the *Logique* sets out and analyzes in the remainder of the chapter referred to above. There is first a division according to the certainty or probability of the sign ("respiration is a sure sign of life in animals"; "pallor is only a probable sign of pregnancy in women"), which is followed by a division according to the copresence of the sign and the signified or their separateness (symptoms of a disease are signs connected with the disease; the sacrifices of the Ancient Law are signs separated from the immolation of Christ). Third, "Signs may be divided into those that are natural, which do not depend on the whim of men, as an image in a mirror is a natural sign of the person represented; and those that exist only from institution and establishment, and which have either some distant relation with the thing signified, or none at all. Thus words and written characters of words are institutional signs of thoughts." Between a natural sign and the thing signified the relation is obvious, and we therefore tend to treat a natural sign as the thing itself even though we are aware that it can be the thing only in a figurative sense. "Without preface or ceremony we will say of a portrait of Caesar that it is Caesar, and of a map of Italy that it is Italy."[13] Some people, however, continue to take this liberty even in the case of the institutional sign, thinking, quite incorrectly, that since signs are often said to be the thing signified, an unreasonable proposition can be made reasonable by being taken figuratively. But by following this rule it would be impossible to produce an extravagant statement: "in fact, the more impossible the literal sense of a proposition, the more easily one would fall back into the sense of sign, which is entirely inappropriate. For who would allow, without preparation and by virtue of nothing more than a private intention, that the sea is the sky, that the earth is the moon, that a tree is a king? Is there anyone unable to see that the introduction of such language is the shortest path to a reputation of madness?"[14]

In reflecting upon this analysis of natural and institutional signs, it should be noted that neither is in a position of subservience to the other. Both deploy different sorts of signification. While the institutional sign pertains specifically to language, that is not the case at all with the natural sign. The domain of the natural sign is the object, not as regards its being or reality, but as regards its activity in a certain capacity, namely, as a representation of another object: indeed, considering the examples, one wants to go further than representation and say as an index or a revelation. The extent of the natural sign is the extent of one object to indicate or reveal another.

But eighteenth-century interest in signs shifts entirely to their constitutive role in language, and while the natural sign is not dismissed, its definition is consequently modified towards that end; no longer does it escape the concerns of language. For this emphasis Condillac is largely responsible. Between Locke and Condillac, thought and language become equivalent through the commensurability of the sign. And it is through the alignment of signs exclusively with linguistic concerns that it is left up to them to explain the genesis and function of thought. Here is to be found the explanation for the well-known interest in the origin of language; in fact, all these researches were necessitated by the conception of the sign. Responsible now for generating and presenting thought, in all its spirituality and undivided unity, through the sequential materiality of sound and written characters – how could these apparently disparate activities be resolved? A perplexing question arising from the fact that language was itself an assemblage of diverse constituents: "speech consists in rendering vocal what the soul has received through the senses, in representing exteriorly what is within and which originally came there from the outside .... How are the real being, the idea, the sound and the letter, four things so naturally opposed and so seemingly irreconcilable, thus brought together?"[15] It was Condillac who determined this to be the proper formulation of the problem; and the solution he proposed, although not definitive (precisely *because* it was not), inaugurated by its example the countless studies on language origins.

The intent of these studies is not to be misunderstood. Interest in the historical origins of languages, in older languages, in mother-languages, this was a pursuit accepted in the sixteenth century and destined to reappear in the nineteenth, but was of no consequence in the period under consideration here (allowing, of course, for occasional vestigial pockets and a certain number of orthodox apologists). Demonstrations of chronological filiation, of the historicity of languages – their moments of birth, their capacity to engender one another, their temporal designations of a place in a historical series – were all of little interest. Condillac was not concerned with language's historical roots and development, but rather with its theoretical necessity and with establishing the theoretical possibility of its origins. And why that? Precisely as a means of grasping the origins and progression of *thought*. For the genesis of ideas now had to be studied in language; and the universality of the operations of human understanding, the immutability of the rules governing thought and grammar in all times and all places, rendered indifferent the historical epoch or civilization into which one projected oneself.

Such historical and cultural variables only added up to the vicissitudes of *grammaires particulières*, inconsequential to the fundamentals sought in *grammaire génèrale*. History is inconsequential; it serves only as a useful backdrop; it is fictionalized without compunction to set the stage. Recall Rousseau's words from his *Discours sur l'origine de l'inégalité*, which itself offers reflections upon the origin of language much influenced by Condillac:

> The investigations into which this account enters need not be taken for historical truths, but only for hypothetical and conditional observations, better suited to clarifying the nature of things than showing their true origin . . . . It is by no means a light undertaking to properly distinguish between what is original and what artificial in the present nature of man, or to form a true idea of a state that no longer exists, perhaps never did exist, and probably never will exist; but of which it is nevertheless necessary to have accurate ideas in order to form a sound judgment of our present state.[16]

History serves to establish distance so that one can construct in it convenient models with selected and controllable variables; it serves, that is, in a role identical to the model of the statue-man, and to the same end: the exposure of the genesis and activities of thought to analysis.

Consider the organization and argument of that most influential work of Condillac's, his *Essai sur l'origine des connaissances*.[17] The importance of

Locke's *Essay* here is indisputable: not without reason did Thomas Nugent translate the *Essai* under the subtitle, "A Supplement to Mr. Locke's Essay on the Human Understanding." Still, it was no simple extension of Locke's thought, for the *Essai* shifts the focus from ideas to signs, from the operations of thought to the effects of language. Not so great a shift as all that, one might think, since Locke also gave some consideration to words and signs.[18] At the end of Book II, for instance, Locke acknowledged the need for a general analysis of words and language, having recognized "that there is so close a connexion between ideas and WORDS, and our abstract ideas and general words have so constant a relation to one another, that it is impossible to speak clearly and distinctly of our knowledge, which all consists in propositions, without considering, first, the nature, use and signification of Language".[19] But he came to the recognition late: "I must confess, then, that, when I first began this Discourse of the Understanding, and a good while after, I had not the least thought that any consideration of words was at all necessary to it."[20] And then he failed to follow it out, or did so, rather, in a misguided way. For his examination of the relation of words to ideas led him to distinguish the clarity of ideas from "the cheat of words", and to conclude that ideas could be detached from language, thereby cleansing them of error and confusion.[21] Signs are thus inessential.

It is this "small mistake" that lies at the heart of Condillac's objection. Locke failed to see "the necessity of signs in the habitual acquisition of the operations of the mind. He assumes that the mind makes mental propositions, in which it joins or separates ideas without the intervention of words. He even pretends that the best way to arrive at knowledge would be to consider the ideas in themselves, although he observes that . . . this is seldom practiced."[22] The project Condillac set for himself in the *Essai* – his improvement on Locke – was to expose simultaneously "the progression of mental operations as well as that of language" – a task that was to be accomplished by an inquiry into the use of signs, "the principle which unfolds all our ideas . . . ".[23] The originality of the *Essai*, the proposition Condillac believed he was the first to formulate, was that "good sense, the mind, reason and their opposites are all born from a single principle, which is the connection of ideas with one another; that, going even further, this connection is seen to be produced by the use of signs; and that, consequently, the progress of the human mind depends entirely upon the skill with which language is acquired".[24] Signs are central. This is stated in the Introduction ("Ideas are connected with signs, and it is by this means alone, as I shall prove, that they are connected with one another"), and repeated in the last chapter ("Thus it has been demonstrated that the origin and progression of our ideas depends entirely on the manner in which we make use of signs"). Our main objective – on this point we should make no mistake – remains "the study of the human mind", an investigation of the source of our ideas. "We must ascend to the origin of our ideas, we must disclose their generation . . . ".[25] But the way to these origins is now at least as much through an analysis of language as through the more typical analysis of the operations of the mind. For that reason the work divides into two parts, the first being "On the materials of our knowledge, and particularly on the operations of the soul", the second "On language and method": it is so divided precisely "in order to complete this *double* object" of tracing the origin of ideas both through the primary operations of the mind and the first stammerings of language.[26] Which is to say that, for all the apparent difference in the subject-matter of the two parts, their purpose is identical.

But Condillac does not wait until the second part to analyze signs; their critical importance (to the soul's operations) is established in the early chapters of the first. In a chapter on the use of signs,[27] three kinds of signs are distinguished – those listed by d'Alembert. The addition of "accidental" signs to the *Logique*'s division of natural and institutional is in order to assure a difference in kind between those signs that are "arbitrary" due to the fortuitous designations of circumstance, and those that are "arbitrary" in that they are artificial, the sound-signs bearing no natural relation to the idea they represent, but that are nonetheless consciously *chosen*, are motivated, in some way involve the processes of reflective thought.[28] As for natural signs, they no longer perform in the same capacity; their function is now entirely generative; as gestures and cries of grief and passion they mark the well-spring and possibility of the higher operations of thought and of language.

The first two categories of signs are sufficient for the limited operation of imagination and reminiscence, but insufficient for the inauguration of memory. This means a particular perception can be renewed and, if reintroduced through a repeated circumstance, it will be sensed that it was experienced before. But the mind cannot remember or recall it at will. Without the institutional signs of language a creature will not be directed by memory (I am hungry; I have been hungry before; I will prepare for the recurrence of hunger), but driven by sensations of the moment (hunger; misery; anguished cry). The vast advantage brought by institutional signs is that they have the power to create memory, for what memory recalls is never ideas themselves – whole, undistilled, uncategorized – but the *signs* of ideas. Our command of memory begins only when command has been established over the signs it revives. Hence, "as soon as a man comes to connect ideas with signs which he has chosen, his memory is formed".[29] "A single arbitrary sign is sufficient to enable a person to revive an idea by himself . . . ".[30] Such a capacity to recall and reflect upon an idea *at will* constitutes "the first and smallest degree of memory"; it marks the beginning of reflection as well as the first formation of the mind.[31] "Let no one object that before this intercourse [between signs and ideas] the mind already has ideas since it has perceptions: for perceptions that never have been the object of reflection, *are not properly ideas*."[32]

Out of natural signs (unreflective cries, grimaces, gestures) Condillac, as is well known, constructed a "language of action". He then showed how the same signs, through a gradual sliding from the spontaneous into the instituted (reactive cry of hunger or of fear sounded willfully signifies desire to eat or danger), could over an extended period produce from this language of action the complex network of invented signs making up conventional language. Rousseau's criticism, in his *Discours sur l'origine de l'inégalité*, is often raised against Condillac – that a language instituted according to convention is an impossibility as it demands an already formed and speaking society, one able to discuss and agree.[33] Language must therefore be given, not created. "In fact", Foucault points out, "the language of action confirms this necessity and renders this hypothesis futile".[34] For the language of action is already transitional, not something given from the start, something entirely natural. Nor is it imitative: a cry does not resemble fear, an outstretched hand hunger. All these natural cries and gestures, why they are signs at all is quite simply because they are the material out of which the first language of action is made, and out of this material because these particular noises and actions have the virtue of being universal reactions, identical among all men. Natural signs are universally imposed. The same cry and the same grimace that has accompanied our own passion or anguish we hear emitted from another, we read upon the face of another. And we recognize it. It no longer stands as simple commotion on the part of a companion but as a representation, a sign. Comprehension begins, coincident with the denotation of the sign.[35]

Thus we witness the diverse constituents of the origin of language: an element of universality, assuring the immutable principles of *grammaire gênérale* and of the operations of the soul; an element of inevitability – in the expression of needs, fears and emotions – which must repeatedly and forever stand ready, if civilization were lost, to activate the capacities of language and of thought; and an element of chance, which, due to the variations of climate and circumstance, channels primitive language into the diverse languages of particular peoples. Chance, for the derivation of a language follows no internal law, traces no evolutionary development, but marks rather, in the passage of time, only the interplay of accidental erosions and incorporations through the outcome of wars, disasters, new governments, commerce, travels, and so on. The formation of Greek, for instance, can be attributed to "Phoenician merchants, adventurers from Phrygia, from Macedonia and Illyria, Galatians, Scythians, and bands of exiles or fugitives who loaded the first stratum of the Greek language with so many kinds of innumerable particles and so many dialects."[36] Nothing, in effect, could be emptier of meaning to a theory of language founded on a system of arbitrarily selected representative signs than to undertake a quest, as was done before this period and would be done in the century following, for the origin of language through derivations traced from the bosom of a mother-tongue. Because no primordial intention was seen to be working itself out in the derivation of a language, no mechanism or force or destiny oversaw and regulated its successive forms – other than perhaps the rule of use defined in the *Encyclopédie* article "Langue": "a language is the ensemble of uses by which the thoughts of a nation are verbally expressed .... Usage is thus the natural legislator of languages". Derivation only marks the sum of accumulated responses to encounters and conflicts, a worn vestige of the arbitrary impositions of events.

The compiling, sorting and arranging of languages undertaken in comparing particular languages sought, by means of these classifications, to shed light on a different level altogether, the level at which *grammaire générale* unveiled its meaningful and immutable principles. Within the context of the sign system laid down in the *Logique*, the search for the origin of language became an attempt to determine the moment of language within the context of all that it represented for human psychology, for rational thought. And in this sense the analysis of origins held no privileged place among the various other sort of eighteenth-century analyses of language. It represented neither an elevation of empirical interests in observation – as distinct from, or as the result of, disenchantment with biblical and rational interpretations of the origin of language – nor some sort of anachronistic historicism. It was simply one more analysis, which could just as well have, and on other occasions did, take the form of a speculative anthropology (savage), a speculative child psychology (proceeding from prelinguistic and prereflective stages to speaking and thinking), or a fictional psychology of the statue-model (again proceeding from prereflective to reflective, but subject to better controls than the child, for purposes of analysis).[37] Like these other analyses, that of the origin of language sought to unfold simultaneously the operations of the mind, the activities of language and the use of signs in a series of enchained propositions that would establish the uninterrupted connection of the progression of these activities – from the sensible idea to the most abstract reflective notion, from the natural sign to the most indeterminate arbitrary sign, from primordial language to the most metaphysical discourse – in this way revealing the natural order of the concatenation of signs and ideas, and the distribution of the faculties of the soul: in sum, the taxonomy of human understanding.[38]

So when Bonnet, in setting out his own interests in the subject of language, writes in the *Essai analytique*, "The origin of language is not my subject: I must assume that language is introduced and consider its general effects",[39] it does not necessarily mean that his interests differ in any fundamental way from those of Condillac. And in fact they do not. His analysis of the specific significance of language for understanding is very similar to that of Condillac's; and despite his disclaimer there is a sense in which he does pursue the origin of language: he pursues it in the preconditions of language in animals, in the stages of human development, in the effects of speech on his statue-man, in language's bearing upon the first activation of the faculties of thought. Where he does *not* pursue it is in fictitious historical reconstructions such as Condillac's account of the two children stranded in a desert following the tower of Babel; nor does he feel the least compulsion to entangle himself in the problems (equally imaginative) of reconciling such speculations on genesis with the pronouncements of *Genesis*. What he *does* try to do is to determine that first indications and the general effects of language while refusing to speculate, at least in his published psychological writings, on its primordial status – whether it was received or made by man, or whether it shares elements of both.[40] And the reason, quite probably, is that it mattered little to his analysis.

This refusal did perhaps have an effect, however, on the status Bonnet attributed to natural signs, for he did not, with Condillac, find in them the scattered units of a rudimentary language of action. Following Lockean psychology, Bonnet, like Condillac, held that all the basic material on which the mind operated derived originally from the senses in the form of sensations or perceptions. Consequently the mind does not "create" ideas; it is so completely dependent for its contents upon representations, or extrapolations from representations, of the objective world that it cannot even imagine something that bears absolutely no relation to received impressions. Sensations and perceptions make up the corpus of "sensible ideas".[41] And in Bonnet's schema of the natural-conventional sign system, the purview of the natural sign coincided precisely with that of the sensible idea: "The sensible idea is a *natural* sign of the object"; "[Sensible] ideas are invested with only *natural* signs, and these signs are the images that objects trace in the brain."[42] Furthermore, the natural sign did not merely coincide with the sensible idea as a received image, it applied as well to the actual movement implanted into the nerve fibers by the object and retained by these fibers in the brain insofar as the original movement could be recalled. As he says in the *Essai de psychologie*: "But we know that certain ideas in the soul correspond constantly to certain movements that objects imprint in the brain. These movements are therefore species of natural signs of the ideas that they excite; and an intelligence capable of observing these movements in the brain would read them like a book."[43] The same view is later affirmed in the *Essai analytique*, in which Bonnet regarded the fibers and their movements "as natural signs of ideas", a careful analysis of which should make it possible "to deduce the order of the generation of ideas in my soul . . . ".[44]

Transferred to a position of identity with the sensible idea, some consequent displacements are produced in the natural sign. The sign does not thereby lose its representative capacity, for the sensible idea *is* the representation of the object.[45] Not as one object representing another (*Logique*) but as an idea representing an object. The sensible idea, both in its physiological and in its spiritual aspects (movement in the fiber, image in the soul), results directly from the action of the object, and yet is something quite different from this action. It is the various transmutations of the action of the object, in its passage through the medium of the body to the soul, that now guarantee both the sign's relation to, yet its distinctness from, the signified. The natural sign, then, is no longer an autonomous index attached to the object it represents; it has been removed to the thickness of the body, where it traces out and sums up – provides the coherence of – all the transformations between object and image. Which brings us to a second displacement: although it has become far more widely disseminated through its identity with the sensible idea, the stature of the natural sign has been diminished by comparison with its place in the *Logique*. By itself it is always insufficient – too determined, too restricted, capable of representing only isolated individual sensations.[46] No longer important for what it reveals, it is valuable only to the extent that it can be used, converted, gotten beyond: "One of the great advantages of artificial signs over natural signs is that the former . . . extend the mind's vision and render it less dependent on sensible ideas."[47] The natural sign has become mechanical, an automatic reaction to the action of the object, its representative image foreordained; it requires no thought, no interpretation; its revelatory power has been replaced with the imprinting of images by rote.

"The natural relation existing between objects and our ideas is independent of the soul. It is not in its power to not be affected by a certain idea when a certain object acts on its senses. The idea is a *natural* sign of the object, and this sign is instituted by the Creator."[48] Moreover, in being joined to the sensible idea, the natural sign has become virtually invisible, seldom discussed despite its range, since it is *assumed* in the discussions of sensations. Previously it had its separate domain; now it is identified with the primitive, with accumulated sensory data – with the raw material from which language and thought derive, but which by itself is always limited. The natural sign has become debased, common, enervated, habitual. Finally, as in Condillac's *Essai*, the natural sign no longer escapes the concerns of language, for language and thought are so intertwined that it cannot, in its identity with the sensible idea, provide the foundation of reflective thought without at the same time providing that of language.

But here natural signs could readily lead back to Condillac's natural cries and to his language of action, for it was known that every sensation is accompanied by feelings of pleasure or displeasure in relation to one's present needs. Gestures and sounds would then indicate unmediated expressions of a creature's response to sensations, and courses of action would indicate volitional responses to preferences generated by sensations. One could therefore envisage a debased language made up entirely of sensible ideas as the language of sentient creatures, a "natural language" where no reflective activity of the mind entered into play. One did not, however, need to imagine children in deserts in order to follow out this idea of a natural language. It could be pursued just as well in speculations on contemporary savages, on men found in the woods, on manlike creatures found in the woods, on men born lacking several of their senses, on prelinguistic infants, on preconscious human models (statue).

Bonnet speculates on natural language in another area as well, in animals, where it appears as "an assemblage of unarticulated sounds uniform in all individuals of the same species, and so linked to the sentiments they express that the same sound never represents two opposed sentiments".[49] As opposed to this natural language, artificial language, of course, constitutes "an assemblage of articulated and arbitrary sounds that have no connection with the ideas they represent other than those given them by institution or convention . . . ".[50] But this being the case, what about those animals, parrots for example, that learn to enunciate words and even sentences of artificial languages? It is a feat demonstrating at least auditory recognition and a vocal facility, but does it indicate as well that they are able to learn artificial languages?

Only the vulgar believe this: "they do not know that to speak is to link one's ideas to arbitrary signs which represent them. The phrases that the parrot repeats with such precision proves nothing except that it has the [sensible] ideas which are attached to the words it pronounces: it would be equally capable of pronouncing the most abstract terms of science."[51] Other domestic animals have been taught visual recognition. They have learned to recognize and arrange alphabetical letters to form words. But again, all that these facts prove is that the brain of an animal, like our own, can form associations of purely sensible ideas.[52]

Although the sort of sensible associations that make it possible for some animals to repeat articulated words is exceptional, sensible associations themselves are not at all uncommon in animals. Animal instinct, as a matter of fact, is nothing other than such associations in direct relation to the preservation and propagation of the animal.[53] Yet instinct, "that imitation of reason",[54] and all associations of sensible ideas, are mechanical and unreflective; the physical ability of animals to mimic conventional languages "is much more mechanical than commonly thought". In fact, it is entirely automatic.[55] Thus, concerning animal language, it can be concluded that

> Beasts have and are able to have only particular or purely sensible ideas. It is impossible for them to be elevated to our universal ideas; this is because they are not endowed with speech. They do not generalize their ideas; they do not form intellectual abstractions. The subject, for them, is confused with its attributes, or rather there is neither subject nor attribute for them. Beings are known to them only through sensible qualities. All their comparisons, all their judgments rest immediately on these sensible qualities. Beasts therefore do not reason, to speak precisely: they lack our mediating ideas because they do not have our signs. Thus, when they appear to reason, they are only comparing or recalling certain sensible ideas, from which certain movements or actions result.[56]

An eventual point is reached at which the analysis of a language of natural signs shifts its focus, becoming less a description of the nature of natural language, more an exposition of its limitations – a demarcation of a condition of deprivation that at the same time serves as as elucidation of the powers of conventional language. For there is a sense in which "natural language" is not natural at all, is unnatural, even a monstrosity. At any rate, it falls into one of those two inclusive categories according to which eighteenth-century teratology determined abnormality: monsters *par defaut*. We have already reached this point in the analysis of animal language, and it is time to examine its limitations, their implications and, then, the fulfillments of true language.

Bonnet presents a summary of these limitations himself, in Chapter VIII of his *Essai de psychologie*, entitled "The state of the soul deprived of the use of speech":

> While man remains deprived of this precious advantage, the sphere of his ideas is confined to very narrow limits. All his perceptions are purely sensible and isolated except for their connection with the circumstances of their origin, or the various relations that resulted from the manner in which they were excited. The ideas are invested with only *natural* signs, and these signs are the images that the objects trace in the brain. The soul is consequently able to recall a certain idea only as long as it is actually occupied with an idea or an image that has a determined relation with this idea. The soul therefore traverses the series of its ideas like a series of pictures. It recalls its perceptions in their natural order or in an order which is nearly the same as that in which they were produced .... Since the succession of these ideas was originally only the succession of movements imprinted into the fibers, as soon as the machine is led to execute one of these movements, it finds itself advancing through the entire series .... The judgments that the soul brings upon objects are not actual judgments: they are only the simple sentiment of the impression of those objects. Every sensation accompanied by pleasure inclines the soul towards the object that is the source of this pleasure: every sensation accompanied by displeasure or pain produces a contrary effect .... All the soul's ideas being particulars or *concretes*, everything being only images – and images of individuals – each idea represents only the object suitable to it, and by itself could be of no assistance in representing analogous objects and of still less use in representing all different kinds of objects. The idea of a man is necessarily the idea of a certain man, of certain traits, of a certain garment, of a certain attitude, etc. Everything here is determined .... Finally, the manner in which the soul deprived of speech expresses its sentiments corresponds completely with the nature of these sentiments or perceptions. These are sounds, cries, movements, gestures, attitudes, etc., which appear to be as connected with the sentiments that represent them as these sentiments are with the objects that excited them.[57]

A description of animals? No, Bonnet has here described the child before it learns to speak; but there is no distinction to be made between its condition and that of the animal. "What I have just said concerning the human soul deprived of speech can be applied to the soul of beasts ... which does not reflect on its actions, which does not generalize its ideas, which is incapable of morality."[58] And in the *Essai analytique* Bonnet's statue is subject to the same state of affairs. There are only those who speak and those who do not, those who think and those who do not, the sentient and the rational.

Two limitations of the natural sign may be singled out: the isolation of sensible ideas and the inability to control these ideas. Without the "precious advantage" of language, control is never established over the sensations with which one is continually bombarded. Instead the sentient being is entirely determined by those sensations, for its soul of necessity responds to every sensation as either pleasant or unpleasant, relative to its momentary needs and wants. Nor can it affect the order of the sensations. "It is the objects themselves that arrange ideas in the brain of the animal."[59] Lacking the ability to arrange and analyze its sensible ideas, the sentient being could be said to exist in a world made up entirely of the unreflected apparent good, amidst an endless series of momentary pleasures, displeasures and pains. This is a vision which may of course cut two ways: one may imagine the idyllic frolics of innocents unencumbered by the duties and sophistications of rational thought; or, considering that the world of the apparent good is in reality the world of the deceptive good, of the concealed bad, of eventual pain, one may well imagine the frustration of invariable disappointment and dissatisfaction: at the end of every false path, like a law of nature, awaits enforced dereliction, degradation, suffering. Beings unable to speak, unable to reflect on themselves or on their actions, unable to generalize their ideas are thereby incapable of morality. But to be amoral, is this to be exonerated from all responsibility, or is it rather to be condemned to suffer, without added insight, all the consequences of wrongdoers? A moral monstrosity.

To this inability to take charge of the order of one's sensations is added the inability of the sentient being to surmount the tyranny of singularity.[60] Every idea of the sentient being is isolated, for it is the great deficiency of the natural sign that it is able to represent only individual ideas. Due to the equivalence of the sensible idea and the natural sign, natural language could never make comparisons and judgments, and the sentient being could never base actions upon judgments. Sensations remain isolated; "judgments" are merely the judging of a sensation to be pleasurable or not. The sentient being, impoverished of language and thought, is confronted with a barrier making impossible the perception of resemblances. And faced with this barrier, which institutes an endless accretion of singularities, nothing is connected or enchained other than that which is connected by the object itself, or by a series of objects in sensible associations. As Bonnet concludes, following a lengthy analysis of the statue's exposure to the fragrances of different flowers: "Each sensation of our automaton is an individual idea, and an individual idea can by itself represent only the same individual. It would therefore be impossible for the statue to be able to acquire general ideas solely with the assistance of the sensations we have caused it to experience."[61]

Language thus becomes a necessity whose functions can hardly be a mystery; it has carved out its place in advance by the weight of its need. It requires no *raison d'être*, only elaboration.

Since humans are mixed beings, activities that affect both their bodies and souls are therefore mixed. Conventional language has this dual aspect, and Bonnet is, in much of his analysis, concerned with the mechanics of language. The arbitrary sign, as received by the senses, is an ordinary sense impression. For language is spoken and it is written: spoken words attach themselves to the auditory nerve fibers in the manner of any other sense impression; the same is true of written words or figures and the optic nerve.[62] That is why parrots can imitate the most abstract terms of metaphysics as readily as the sounds of other animals. Yet there is obviously something special about an arbitrary sign, since it is no more closely connected to one sensation than to another. A *natural* sign of pleasure, for instance, would be associated with a particular agreeable sensation: if the statue preferred the scent of a violet to that of a rose, then the scent of the former would be "a *natural* sign of pleasure for the statue". Thus it is too determined, its representation of pleasure too restricted. But with an arbitrary sign, "the word 'pleasure' can be connected indifferently to all sorts of agreeable sensations . . . . Hence it follows that *the more indeterminate the sign, the more it is sign*; for it has greater representative capacity". And if it should happen that a sign representing a general idea were to recall a particular idea, "it is due to a circumstance absolutely foreign to the sign as *sign*; it is because habit has strongly enchained it to a particular idea".[63]

Language is first acquired by learning to associate the idea of an object with the word representing that object, a task accomplished only by repeatedly hearing the sound pronounced in view of the object. "Once this connection is made, the two ideas are reciprocally recalled: the word becomes a sign of the object; the object occasions the recollection of the word."[64] This association is made in the manner of all sensible associations. If, for instance, the scent of a rose is presented to the statue while the word "rose" is repeated several times, auditory fibers communicate with olfactory fibers and the scent is connected to the word. Two orders of movements, corresponding to one another, are actually introduced into the fibers: that of the imprinting movements of a specific smell and specific sound, and that of a communication between the two orders of fibers. As a result of this double movement, the two orders of fibers contract a "connection of action" with one another by which "they tend to reciprocally communicate impulsions".[65] Once the connection has been made between the idea and the sign, the sign alone is sufficient to bring about the recollection of the object.[66] All these associations and communications between fibers present a challenge to Bonnet's principle of the structural specificity of nerve fibers, one met with the proposal of mediating structures, links that he specifically terms *chaînons*. "Nothing appears to favor this propagation more than the relation of structure and the analogy of elements. One can thus conjecture with some probability that the *chaînon*, which connects two orders of sensible fibers, contains elements analogous to those of each order . . . ".[67] The same mechanics operating between a sensation and its representative sign also operate between sequences of sensations and signs. "Thus it is that we retain a sequence of ideas represented by a sequence of words of a discourse. The *chaînons* that link together the fiber bundles appropriate to these ideas and to their signs create a single chain out of all these bundles along which movement is propagated in a constant order."[68]

As can be seen, this analysis is only an extension of the theory of associated sensible ideas. "Ideas are joined (*s'associent*) to their signs as they are joined to one another. The same mechanism that connects an associated idea to the principal idea connects the sign to the idea that it represents." Yet in this adjoining of ideas to their signs a fundamental shift has occurred. For as Bonnet goes on immediately to say: "This double association of ideas with one another and with their signs constitutes the foundation of the knowledge of each individual."[69]

Several specific physiological answers are in fact offered regarding how the sign, as sensation, establishes a domain from which, as a free agent, it can be connected variously with sensible or with reflective ideas. First, through an extension of his specific fiber theory, Bonnet argues that there are intellectual fibers appropriate to artificial signs, as opposed to the sensible fibers of natural signs; in the same way that sensations are attached to sensible fibers, the soul's reflective ideas adhere to these intellectual fibers, which serve the operations of the understanding. And since artificial signs enter through the senses of hearing and vision, "it can reasonably be supposed that intellectual fibers are only an extension or continuation of those serving sight and sound".[70] Second, a distinction is made between the faculty which recalls sensible ideas (imagination) and that which recalls intellectual ideas and "words representative of things" (memory).[71] Finally, an analysis is undertaken of the special links (*chaînons*) joining arbitrary signs to sensible ideas, to general ideas and to one another.[72]

The physiology of signs is of course not the whole of speech, but it did constitute a special interest of Bonnet's, and in it he situated one of the fulfillments of language. For by means of the mechanical "double association of ideas with one another and with their signs", connections are established between ideas as well as between an idea and its representative sign. Ideas are numerically increased since relations perceived between ideas, as a result of these connections, are themselves ideas; and the ideas are more orderly since the links joining ideas to their signs "create a single chain" of associations.[73]

This leads to a second fulfillment: singularity is overcome. It is not that the employment of signs constitutes the source of abstraction – a function attributed to the attention – but their employment vastly facilitates the ability of the attention to separate ideas and to fix the ideas separated, since terms are made available to represent the various processes and products of abstraction.[74] Without signs to represent what has been abstracted, the limited capacity of the attention becomes divided and weak; but, with the assistance of signs, the mind can for instance abstract the figure of an object and represent this figure by lines: "its attention will be concentrated in this figure because the abstract idea will exist *apart*. It is this sort of abstraction that is the object of geometry. The object of geometry does not exist in nature".[75] The barrier imposed upon natural language is that beyond which the process of abstracting cannot proceed without the assistance of arbitrary igns. With them, the force of attention remains focused and abstractions are carried further: "The ease of separating or of abstracting leads to the *generalization* of the ideas that have been abstracted."[76] In generalization the mind does not simply detach a component from the concrete idea: instead, it detaches all the characteristics of its individuality in order to form the general idea of the object.[77] Through the consideration of a particular oak, the rational soul "is elevated by degrees to the general ideas of plant, of organized body, of body in general, of being".[78] The general idea "retains nothing at all of the particular. The characters that it contains are thus in equal accord with all oaks; for they are the expression of what is in all oaks".[79] It is an extrapolation made by the mind, an idea without an archetype outside of the mind, since "No oaks exist in general".[80] Generalizations are intellectual abstractions rather than sensible abstractions, and the products of generalizing are called general ideas or notions. In these general ideas the solution to singularities is found. As Bonnet says: "It is thus in extending and in facilitating the exercise of the attention that the use of arbitrary signs gives the soul the means of decomposing and knowing *the general relations of resemblance* joining the beings of a single species, a single genus, a single class."[81] And the perception of these resemblances requires notions both of singularity and of generality – for distributions can result only from the double play of identity and difference.

If the mind considers a concrete object in relation to its *individuality*; if it designates with terms the particularities that it discovers there and that characterize it as *individual*, the mind will acquire the *particular* notion of this object, and the expression of this notion will be a description.

If the mind considers the object in relation to the objects that resemble it; if it expresses the object in terms of what these objects have in common, it will acquire the *general* notion of the object; and the expression of this notion will be a *definition*.[82]

And then a final fulfillment of language, closely connected with the other two: it gives man "the most absolute control over his ideas".[83] Language specifically activates the powers of reflection, the operation by means of which the mind acquires general ideas,[84] for "this power [of reflection] presupposes the use of signs by which we generalize our ideas".[85] Man thus gains command over his thought; and unlike those beings without language whose ideas are arranged by the objects themselves, it is by means of this control that man is able to arrange his thoughts as he desires, according to the goals he proposes for himself.[86] This ability, says Bonnet, is what is called *method*.

The special status accorded "analysis" in the works of Bonnet, and in those of other eighteenth-century figures, can thus be appreciated. The incorporation of analytic method into the very structure of language was not artificial in the least, but derived directly from the conception of language and its functions. For analysis – the proper method of ordering and arranging through the decomposition and then reassembly of a whole as the means of its appropriation – coincides precisely with the operations and concerns of language and thought. If it is only through the imposition of one's own arrangements and distributions upon things that relations are perceived, that knowledge is appropriated, then analysis is the method of this arranging and appropriating, the means by which language and the understanding spread the canopy of resemblance over the field of singularities. Its universalizing tendency is no more than a reflection of that of language and thought.

But analysis is not a methodological ideal, something set apart, something parallel, upon which language should pattern itself if it is to conduct itself properly. The reverse is the case. Language exemplifies analysis by its very existence; it testifies to the "naturalness" of analytic method by its mode of operation. Specifically, the same sign system that serves to designate ideas and to communicate thought – that is in general responsible for linking knowledge to language – also assures the correct performance of logical operations. "The science of language and the study of languages is necessarily based upon the analysis of thought. Each of our particular thoughts is a whole, an ensemble." And each requires the decomposition of analysis in order to be grasped in the combinatory variety of its discrete elements. Consequently, "In order to enunciate thought it is necessary to analyze." So writes Ferdinand Brunot of eighteenth-century language.[87] But the point is even more clearly expressed in the *Encyclopédie* article "Grammaire": "With the help of abstraction, logic follows and analyzes thought, entirely indivisible as it is, by separately considering the different ideas by which it makes up the object, and the relation that the mind apperceives between them. It is this analysis which is the immediate object of speech; and it is for this reason that the art of analyzing thought is the primary foundation of the art of speaking, or in other words, that a sane logic is the foundation of grammar." And the point is made earlier in Condillac's *Essai*, even built into the structure of the work, for the second part is made up of two sections: one on language followed by one on analytic method, set forth as the "most natural order". It is in the first part, however, that analysis is defined: it "consists only in the composition and decomposition of our ideas, in order to compare them differently, and by this means to discover the relations existing between them, together with the new ideas they are capable of producing". It is "a sort of calculation (*calcul*)", which "presents ideas only a few at a time, and always in the most simple gradation".[88] Now if this is compared with the account, coming a little later, of the advantage of signs, we find striking similarities. Only through the use of signs are we able to reflect on the elements of complex ideas, "which, by being combined, make us envisage them as though they were only one idea". "We conclude that in order to have ideas on which we are able to reflect, we need to imagine signs serving as links to the different sets of simple ideas."[89]

Bonnet's views on analysis only repeat those just expressed. When he speaks of "method", he means analysis, and language gives us this method which puts us in control of our ideas. More generally, the very study of psychology, like that of nature, is composed of two related parts, "the historic part and the systematic part. The first contains the exposition of the facts; the second, their explication". The method of explication? " . . . this method is analysis." "One must atomize each fact, decompose it into its smallest parts, and separately examine all these parts. One must seek the relations linking these things to one another and to analogous things, and find results which can become principles." With regard to the *Essai analytique*'s analysis of the activities of the mind, Bonnet writes: "I would not say I have attempted to enchain all the propositions to one another: it would be more correct to say that they are enchained to one another of themselves. I have only had to follow the analytic thread I had in sight."[90]

Analysis is not external or artificial; it speaks from within the system of signs, from within its order of representative designations, offering up its decompositions and distributions. The connections between propositions and the relations between thoughts emerge from the very structure and activity of language. That is why "the art of analyzing thought is the primary foundation of the art of speaking"; that is why, at the foundation of *grammaire général*, a "sane logic" rules. For language, in its dissemination of arbitrary signs, is analytic. It traverses unitary and composite representations, fracturing them, naming them and presenting them part by part – or several parts together – for observation and comparison. And language, in its grammar, in its system of word structures (alphabetical letters, morphemes) and word arrangements in sentences (syntax), is sequential. It assembles all the scattered and isolated ideas of sentient thought, and it reassembles its own analytic scissions into the linear sequence of unfolding discourse.

And so, two axes of language may be characterized. On the one hand, the associative function of signs, which increases the connections between ideas. Its spatial image is that of a web, a growing network of interconnections; its aspect, physiological. On the other hand, the distributive function of signs, which derives from the perceived relations of resemblance and which establishes control over sensible ideas in the form of serial arrangements by means of analytic discourse. Its spatial image is linear, that of a chain, a chain of ideas; its aspect, intellectual, the activities of the understanding enhanced through the instrument of language and carried out through the method of analysis.

A few concluding remarks. The deeper concern of language analysis, then, was not with the origin of signs in their visible and audible manifestations as articulated sound or as written characters, but with the powers that accompanied the acquisition of these signs, the connections, the generalizations, the methodological control of ideas. Where is the moment of this acquisition? The moment of control that is simultaneously the moment of self-control, of the awareness of oneself as distinct from one's perception (the concurrent moment of reflection and self-reflection); the moment when words are no longer parroted, but tied to the ideas they represent, when they become meaningful; the moment of truly *speaking*, out of one's understanding and self-understanding? But does not the assiduousness with which all these analyses were conducted regarding this common end suggest a pressing concern? Why all the attention directed here? Herbert Josephs, in his book on Diderot, makes this point: "Neither Condillac, nor Rousseau, nor, in fact, any French thinker of the Enlightenment was to succeed in bridging the gap between gesture and language and extricating language from an arbitrary relationship to reason."[91]

The displacements of the natural sign, brought about as a result of joining the sign system to sensational psychology, have already been noted. But it must now be added that its displacement did not leave the artificial sign unaffected. Rather, the relocation of the former was accompanied by the more or less permanent dislocation of the latter; a permanent imbalance was instituted within language which kept it in search of its own possibility. If the domain of the natural sign became enclosed within that of the sensible idea, the domain of the arbitrary sign remained unsettled.

On the one hand, the arbitrary sign has difficulty demonstrating its *difference* from the natural sign, a problem that arises for Bonnet particularly in his attempts to situate the sign physiologically. How does the sign, which enters as a purely sensible idea, acquire its special status? Bonnet's answer we have seen: a special category of fibers (intellectual), a special apparatus of recall (memory), and a special terminology for the discussion of fiber connections (*chaînons*). But intellectual fibers, as it turns out, are only sensible fibers serving a special function, indistinguishable except by means of the signs attached to them: they are only "an extension or continuation of those [fibers] that serve vision and hearing".[92] With memory and imagination, the distinction again dissolves when one finds that they operate by the same mechanism, that they "are actually only the same faculty considered under different faces".[93] Even as used by Bonnet they sometimes become equivalent: "I call this faculty that preserves received impressions *imagination* or *memory*."[94] Finally, *chaînons* offer no indication that they contain a special facility for making connections; they offer no physiological explanation for the way in which arbitrary signs are able to establish so many connections between fibers: rather, the term is extended to include all fiber connections. In truth, every distinction introduced into the physiological explanation of arbitrary signs in order to establish their separate sphere fails.

On the other hand, there is a problem of demonstrating *identity* with the natural sign. The more indeterminate the arbitrary sign – the more it fulfills its capacity as sign, that is – the more indeterminate its relation to the signified. This formed the basis of Diderot's mistrust of language, which was directed specifically at its adequacy of representation. "Diderot contended that there was no necessary or even valid relationship between words and the ideas they have been made to represent." Moreover, language abstracts, and abstraction is itself the source of *error* in philosophical thought. Abstraction, which increases along with the civilizing process, widens the breach between sign and idea, and it can be narrowed again only by remaining attached to sensory objects. But words have come to replace concrete representations. The words and sense impressions that are inseparable in childhood dissolve with age, leaving empty abstract forms in the place of what were once concrete representations. Similarly, with the advance of civilization language has become more abstract, more detached, more mechanical, more impoverished: "all cultivated languages must . . . lose their vigor as they increase in clarity".[95]

This represents a critique of Condillac through a return to Locke's position that the imposition of signs distorts pristine thought, that language is not the guarantor of method but the bearer of error. But it does not necessarily return to the view of Locke's most directly criticized by Condillac – that we can think independently of language. Which leads to the more melancholy conclusion that, although language distorts, we are speaking beings and cannot entirely escape the kinds of errors it produces. Diderot was not alone in voicing this concern, nor even the first; quite possibly he based his arguments on Maupertuis' *Reflexions sur l'origine des langues et la signification des mots*, published in 1748, or only two years after the *Essai*. From Maupertuis' perspective, languages do not express primary ideas so much as they do base prejudices; they are seen to be made up of words passed down from simpler and cruder times, and these original designations of ideas have consequently affected all our thought. Yet we are bound to language and what knowledge we have of things is determined by language. We too readily assume that language is an advantage, that through it thought is clarified, that we benefit from its decompositions. But these decompositions are deceptive: "we misunderstand our result: we have taken each of the parts for things; these things we have then combined with one another in order to uncover expedient or contrasting relations, and out of all this is born what we call 'our sciences'."[96]

On this matter, Bonnet stands on the side of Condillac. He is concerned not with the errors brought by language but the benefits. Not unrelated to the problem of identity, however, is a revision he makes in the *Essai analytique* when he extends universal abstraction to sentient beings, whereas earlier, in the *Essai de psychologie*, it had been denied them; for he is attempting to shorten the gap between the abstracting capacity of sentient beings and the generalizing capacity of rational beings, making the difference one of degree rather than of kind.[97] But the truth is that the powers of universal abstraction are devalued even as they are granted to sentient beings, for these beings are still incapable of overcoming singularities because that capacity has now been advanced to generalization. It has been extended for a different reason: to diminish as far as possible the break in the continuum from sensible abstractions to the abstractions of rational beings. The break actually disappears altogether through the intervening agency of language; by means of arbitrary signs, the same faculty (attention) carries out the same operation (abstracting) to a higher degree (generalization): "It is thus in extending and in facilitating the exercise of the attention that the use of arbitrary signs gives the soul the means of decomposing and of knowing the general relations of resemblance that join the beings of a single species, a single genus, a single class."[98] But it is difficult to hide the fact that generalizing is a different activity from abstracting, as Bonnet defined them. And so the problem presents itself: conventional language and intellectual abstractions remain disconnected from natural signs and sensible abstractions.

The relation between natural and conventional language, between sensible and general ideas, is thus never entirely secure. And the instability is on the side of the artificial sign, encumbered with the responsibility of representing in such a way as to demonstrate both an identity with and difference from the natural sign. If this did not lead to a crisis of confidence in rational thought, it did engender nagging concerns and persistent efforts to determine precisely the nature of the relation between natural and conventional language.

A potential challenge to rational thought was thus diverted into language theory. If rational thought no longer felt compelled to defend itself, as it had in the seventeenth century against skeptical attack,[99] there was still a need to grasp its ability to convert singularities into generalities, isolated impressions into orderly distributions, in relation to human interests and needs. And the burden of this explanation fell upon the representative capacity of language, which had to be at once joined to the sensible idea yet free of it, a representation of the most abstract idea yet tied to its origin in the sensation. It was a difficult position to maintain, there being, as we have seen, a potential collapse of the arbitrary sign into the natural sign on the one hand; and on the other, a complete break with it, language itself threatening to become a purely indeterminate sign, a representation of abstract thought unable to demonstrate its connection with the particular, the individual, the sensible.

Thus the predicament of language and, consequently, the reason for so many studies of its origin and acquisition: it seeks an origin, a basis, a meaning where there can be none – for the sign is arbitrary and, as sign, entirely indeterminate; it cannot reveal meaning, not even in the very moment of its designation, nor can it be gotten behind – and yet where there must be one. For otherwise the generalized idea of reflective thought becomes as arbitrary as the sign by which it is represented. It is quite simply the question, as Foucault has said, which from the seventeenth century asked "how a sign could be linked to what it signified", and whose answer in the eighteenth century inevitably came by way of analyses of representation.

NOTES

[1] Cf. Foucault, *The Order of Things*, 78–9.

[2] Bonnet nowhere explicitly discussed sign theory in the sense of questioning its theoretical status. Like everyone else he simply assumed and built on the definitions given in the *Logique de Port-Royal*, a work with which he was very familiar. In his *Mémoires* he claims to have read and reread it often.

[3] On this point see the Introduction to *L'encyclopédie "Grammaire" et "Langue" au XVIIIe siècle*, ed. and intro. Sylvain Auroux (Paris, 1973). The etymology is that of Charles de Brosses, *Traité de la méchanique des langues et des principes physiques de l'étymologie* (Paris, 1801).

[4] Antoine Arnauld, *The Art of Thinking: Port-Royal Logic*, trans. James Dickoff and Patricia James (New York, 1964).

[5] *Logique de Port-Royal*, critical ed. by Pierre Clair and Francois Girbal (Paris, 1965). This passage is inadequately rendered in the translated edition; and in *The Order of Things*, the last phrase has been mistranslated to read, "and its nature consists in exciting the first by means of the second" (63–4), rendering Foucault's subsequent discussion incomprehensible.

[6] *Ibid*.

[7] *Encyclopédie*, article on "Grammaire". See also Hans Aarsleff, *The Study of Language in England, 1780–1860* (Princeton, 1967), 15.

[8] Michel Foucault, "La grammaire générale de Port-Royal", *Langages*, 7, September, 1967, 7.

[9] *Ibid*., 8.

[10] *Encyclopédie*, article on "Grammaire".

[11] Foucault, *The Order of Things*, 42, 63.

[12] *Ibid*., 42–3.

[13] *Logique de Port-Royal*, part II, Chapter XIV.

[14] *Ibid*.

[15] De Brosses, *Traité*, I, 1.

[16] Rousseau, *Discours sur l'origine de l'inégalité*, Introduction. See also Jacques Derrida, *Of Grammatology*, trans. Gayatri Chakravorty Spivak (Baltimore, 1976), 192: "The *Discours*, which begins by 'setting aside all facts' in order to describe an ideal structure or an ideal genesis, is in [Espinas'] view incompatible with the *Essai* [*sur l'origine des langues*], which calls to some extent upon Genesis, alludes to names like Adam, Cain, Noah, and deals with a certain factual content which belongs to history as well as to myth. One must of course study carefully Rousseau's usage of this factual content, and examine if, by using it as a guide to reading or to pivotal examples, Rousseau does not neutralize it as fact – a step which he permits himself to take in the *Discours* . . . ". For supporting material regarding the views expressed in this paragraph, see Foucault, *The Order of Things*, esp. 89–90 and 104–9; and Hans Aarsleff, "The History of

Linguistics and Professor Chomsky," *Language*, **46**: 3 (1970), 577–8, and his "The Tradition of Condillac," *Studies in the History of Linguistics*, ed. Dell Hymes (Bloomington, 1974), 107–11. In both places Aarsleff points out that misunderstanding of the eighteenth century's use of origin questions began with the nineteenth century, which had so generally replaced the view of a universal human nature with that of a nature constituted through time – where an historical perspective so dominated – "that it could no longer see the attempt to deal with origins in any other light". Needless to say, many studies of more recent date than the nineteenth century have persisted in its misconceptions.

17 Aarsleff says, "The truly creative period [of the origin of language debate] was intense and brief. It began in 1746 with the publication of Condillac's *Essai* and it ended twenty-five years later in the final month of 1770 when Herder wrote his prize-essay. ... With a singleness of origin that is rare in the history of ideas, the fountainhead of this debate was the *Essai*, which in turn drew its inspiration from Locke's *Essay*." "The Tradition of Condillac", 94.

18 He did so at least in part, and perhaps fully, under the specific example of the *Logique*. Without doubt Locke knew Arnauld's work. Of the two concepts of ideas found in his *Essay* – ideas as mental operations, and ideas as objects of the mind – the latter is taken from the *Logique*. See John Yolton, *The Locke Reader* (London, 1977), 5, 110. So also are the ideas found in the *Essay* regarding logic (a subject Locke is generally thought to have despised) and probability. And there is now some fairly recent evidence that Locke even extensively involved himself in a translation of the *Logique*. See the foreword to *The Art of Thinking*, by Charles W. Hendel, xix note. Also see John Harrison and Peter Laslett, *The Library of John Locke*, The Oxford Bibliographical Society, n.s. XIII (Oxford, 1965), 75. Locke's acquaintance with Pierre Nicole (a collaborator on the *Logique*) is well known. In 1676, during his residence in France (1675–9), Locke translated three essays that Nicole had published in 1671. In one of these essays Nicole says, "Knowledge is either of words, or of things, or of actions." For this see J. W. Yolton, *Locke and the Compass of Human Understanding* (Cambridge, 1970), 1; also see Maurice Cranston, *John Locke: A Biography* (London, 1957), Chapter 13.

19 John Locke, *An Essay Concerning Human Understanding*, Alexander Campbell Fraser (ed.) (New York, 1959), book II, Chapter XXXIII, § 19.

20 *Ibid.*, book III, Chapter IX, § 21. Although, several weeks after Locke's return from France (in April 1679), he wrote his friend Thoynard that his "book [the *Essay*] was completed", he nevertheless held onto it for the next ten years, making numerous additions and revisions. It was in this period that he grew increasingly concerned with the problem of obscure and confused ideas (again influenced by the *Logique*), finding the nexus of the problem in the unreliability of words. See Frazer's Prolegomena to the *Essay*, xxviii–xvix.

21 *Ibid.*, book. IV, Chapter V *passim*, and Chapter VI, § 1.

22 Étienne Bonnot de Condillac, *Essai sur l'origine des connaissances humaines* in *Oeuvres complètes*, new ed. (Paris, 1827), I, part I, Section IV, Chapter 5, § 4.

23 *Ibid.*, Introduction.

24 Quoted in Aarsleff, "The Tradition of Condillac", 96. My translation.

25 *Ibid.*, Introduction.

26 *Ibid*. My emphasis.

27 *Ibid.*, part I, Section II, Chapter IV.

[28] Condillac never attempted to reduce *reflection* to something as mechanical and reactive as the recording and transmission of a perception. That interpretation apparently derives from the subtitle to the original French edition of the *Essai* ("A work in which everything concerning the understanding is reduced to a single principle"). The "single principle" is, as I am trying to show here, the acquisition and function of signs. Aarsleff has covered this same issue in his articles, cited above. However, he considers the "single principle" to be the connection of ideas (*liaison des idées*), whereas Condillac says, as I have quoted in the text, that ideas are connected with one another *only by means of* their connection with signs. What Condillac means by this will become clearer as his view of signs is presented. See also Aarsleff's article, "The Eighteenth Century, Including Leibniz", *Current Trends in Linguistics*, Thomas A. Sebeck *et al.* (eds.) (The Hague, 1974) 416–17.

[29] *Essai*, part I, Section II, Chapter IV, § 46.

[30] *Ibid.*, part I, Section II, Chapter V, § 49.

[31] *Ibid.*, also part I, Section IV, Chapter II, § 25.

[32] *Ibid.*

[33] Also see *Encyclopédie*, article on "Language".

[34] *The Order of Things*, 106.

[35] See *ibid.*, 104–7. This follows his argument, except where he claims that cries and gestures making up the material of language of action are not yet signs. They are precisely what Condillac defines natural signs to be, and he does so, I think, to indicate their capacity for signification, in their universality, even before their use as signs. Nature, that is to say, provides the material for signification from the start. The whole employment of the natural sign, in the *Essai*, as opposed to its place in the *Logique*, is as a prelude and prod to the initiation of language and thought.

[36] Abbé Pluche, *La Mécanique des langues*, quoted in *The Order of Things*, 90.

[37] See Bonnet, *Essai analytique*, § § 11–13, where Bonnet says children are too difficult to observe, for they are hardly born before they are exposed to all sorts of impressions, and the chains cannot be followed. "*We thus resort to a fiction*: it will not be natural, but it will be founded in nature" (my emphasis). Its advantage will be that we can follow its development at our leisure.

[38] Cf. *The Order of Things*, 72–3.

[39] *Essai analytique*, § 218.

[40] Bonnet did write several letters reviewing Antoine Court de Gébelin's massive nine-volume work *La Monde primitif*, and in particular those passages in which Court de Gébelin offered his views on the origin of language. Court de Gébelin argues that language was not given by God. What was given was a natural capacity for speech, but man actually acquired artificial language on his own; and when God spoke to man he must have done so in man's own language, using words he would use. It doesn't really matter to Bonnet, since he views Moses' history of the creation, and of Adam and Eve, to be purely allegorical. He also puts forward here some very sketchy speculations on the natural origins of language, but on the whole seems to think the problem unresolvable. For all this, see MS 35–44 – 759, "Lettres de MM. Bonnet et Court de Gébelin sur le monde primitif et l'origine des langues". My point is this: if Bonnet showed enough interest in the subject to read the literature and to comment on it in letters, but still stayed away from it in his psychology, it only confirms his statement that he did not consider it relevant to his work.

41 According to Bonnet's definition, sensible ideas may be either sensations or perceptions. Perceptions differ from sensations in that they reflect a lower degree of stimulation of the sensible fibers: perceptions are the simple apprehension of the object; sensations are perceptions accompanied by pain or pleasure.

42 *Essai de psychologie*, Chapter VIII; *Essai analytique*, § 218, and see also § 228. Emphasis is Bonnet's except when specified.

43 *Essai de psychologie*, Introduction.

44 *Essai analytique*, Preface, xxxii.

45 *Ibid.*, § 214.

46 *Ibid.*, § § 830, 839, 228.

47 *Ibid.*, § 266.

48 *Ibid.*, § 218.

49 *Contemplation*, part XII, Chapter XXXII, 333. Numerous observations have demonstrated characteristic patterns to the sounds made by each species. *Ibid.*, 335; *Essai de psychologie*, Chapter XIX.

50 *Contemplation*, part XII, Chapter XXXII.

51 *Ibid.*, Chapter XXXIII.

52 The operation of sensible associations is in fact not different in kind from the creation of concrete ideas, formed when an object acts on several orders of fibers simultaneously due to a compound activity on the part of the object, an activity that is actually an aggregate of particular forces. These forces are reunited in the concrete idea, "which is like the ideal expression of these forces." (*Essai analytique*, § 210) When the concrete idea is reproduced, all the particular ideas constituting it are simultaneously reproduced. How does one account for the communication between the different orders of fibers that concur to produce the concrete idea? It is in answer to this question that an account of the concrete idea corresponds with that of the sensible association of ideas, for Bonnet always tries to put the basis of such communications within the structure and arrangement of the sense fibers. (*Ibid.*, § 214) Fibers develop associations with one another through the repetition of the same concrete sensation, by means of which they gradually contract the habit of reciprocally acting upon one another. Thus, as soon as one or more of these specific fibers is excited, the movement will instantly be communicated to related fibers, "and the total idea will be reproduced". (*Ibid.*, § 651) "Through constant repetition, one contracts the habit of associating, e.g., a series A, B, C, D, E, F as a unit whose parts are arranged in a particular sequential order. What has been said on the reproduction of the ideas that make up a concrete idea must be applied to the reproduction of all concrete ideas that have been excited simultaneously or successively by different objects. The order in which they have been excited or in which they are succeeded will influence their reproduction . . . ". (*Ibid.*, § 215; see also § § 622–51, 684–5, 789–823) In sensible associations, the sense impressions make their own combinations and create their own series.

53 Since the number and the variety of sensations as well as the associations of these sensations in the brain are fundamental in determining the level of development of an animal's instinct, the number of its senses and the ability of these senses to associate sensations will bear directly on the animal's "intelligence". It was on this basis, for instance, that Buffon placed the elephant next to man in the animal kingdom. For though its trunk did not add new varieties of sensations or an increased number of them, it did bring certain types of discrete sensations into relation to one another; it united the

senses of odor and touch; it could measure distances or judge resistance; it was extremely mobile and could act like a long finger or like a powerful arm; and, most important, it combined all these things in the brain. This combinative function alone was believed, by Buffon, to increase the elephant's instinct sufficiently to make it the most intelligent of all animals. Bonnet, on the other hand, emphasized domestication and educability as contributing most to the growth of the association of sensible ideas in animals. Since he preferred the orangoutang over the elephant as the animal closest to man (actually for structural reasons, it being the most perfect prototype of man), he emphasized its susceptibility to domestication and education. Indeed, he exaggerated it, arguing at one point, as quoted in the last chapter, that with a proper education, the orangoutang could make "an adroit *valet-de-chambre*" (*Contemplation*, part II, Chapter XXX, 175–6 note). See *ibid*., part XII, Chapters XLVI, XLVII; *Essai d'application des principes psychologiques*, XV, 157–64. An exaggeration, but on the other hand, provided they could be taught language, la Mettrie in *L'homme machine* was willing to grant them even more privileges: "Then he will no longer be a wild man, nor a defective man, but he would be a *perfect* man, a little gentleman . . . ".

54 *Contemplation*, part XII, Chapter XLVI, 432.

55 *Ibid*., Chapter XXXIII, 350, and 352 note 7.

56 *Ibid*.

57 *Essai de psychologie*, Chapter VIII, 17–21.

58 *Ibid*., 22.

59 *Essai analytique*, § 315.

60 By singularity is meant exactly what was meant in the *Logique*: "Although all existent things are singulars, still, by means of the abstractions just explained, we are able to have several kinds of ideas . . . . Ideas which represent only a single thing are called singular or individual ideas and what they represent, *individuals*. Ideas which represent more than one thing are called universal, common, or general ideas". *Logique de Port-Royal*, part I, Chapter VI, entitled "Ideas considered according to their generality, particularity and singularity".

61 *Essai analtyique*, § § 787–8.

62 *Ibid*., § § 220, 223, 454–5, 851.

63 *Ibid*., § § 828–9. Emphasis mine on the second of the three quotes. As has been said above, the natural sign tends to be invisible, to be buried within the sensation; and when it is specifically discussed, there is a tendency to give emphasis to *natural*, as can be noted in the various quotes above. Wherever the term "sign" appears without qualification, or with such qualifications as "representative sign", the reference is to arbitrary signs – in Bonnet's works and here – unless it is clear that the reference is to natural signs.

64 *Essai de psychologie*, Chapter X.

65 *Essai analytique*, § § 789–804.

66 *Ibid*., § 221.

67 *Ibid*., § 806.

68 *Ibid*., § 810.

69 *Ibid*., § 822.

70 *Ibid*., § 851, 223.

71 *Ibid*., § § 212–13, 215–16, 222–3; *Essai de psychologie*, Chapter VIII.

72 Bonnet used the term *chaînon* for the first time in *Essai analytique,* § § 806–22, and apparently coined it specifically for discussing the various connections of arbitrary

signs. However, even in this discussion he extends the scope of these *chaînons* to include the connections between sensible ideas – perhaps, again, because the connections are analogous, one kind of fiber being only a continuation of the other.

73 *Essai analytique,* § § 217, 810.

74 *Ibid.,* § 225; see also § § 207–9, 226–9, 267–70. Analogous to this is the advantage Bonnet attributes to writing (§ § 813–16), which relieves the mind from the necessity of holding a long series of arguments or a sequence of discourse. It is relieved "of the double task of composing and retaining", which can fatigue the brain fibers since "the force of the intellectual fibers is not infinite". Thus language assists a limited force of the soul (attention) while writing assists a limited capacity of the fibers. Bonnet no doubt had his own condition in mind. As said in the first chapter, though he had damaged his eyes and was seldom able to see well enough to write, he had learned to compose long passages (five to ten pages with regularity, but as many as thirty or forty) and remember them until someone would transcribe them (sometimes for several weeks or more). He composed the entire *Essai analytique* in this way, and was convinced it was a talent born of necessity which most writers would develop given his circumstances.

75 *Ibid.*, § 225.

76 *Ibid.*, § 227.

77 *Ibid.*, § 228.

78 *Ibid.*, § 227.

79 *Ibid.*, § 228.

80 *Ibid.*, § 229.

81 *Ibid*. My emphasis.

82 *Ibid.*, § § 231–3.

83 *Ibid.*, § 314. In *Essai de psychologie*, Chapter VIII, he says, a "most despotic" control. As for the relation of the physiology of signs to the control of ideas, Bonnet says his discussion of *chaînons* sufficiently demonstrates that they are the most important part "of this admirable instrument through which our soul exercises the most absolute mastery over all its ideas" (*Essai analytique,* § 822 note).

84 *Ibid.,* § 259.

85 *Essai d'application*, 162.

86 *Essai analytique,* § § 310, 315.

87 F. Brunot, *Histoire de la langue française des origines à 1900*, VI (*Le XVIIIe siècle*) (Paris, 1932), 904.

88 *Essai*, part I, Section II, Chapter VIII, § 66.

89 *Ibid.*, part I, Section IV, Chapter I, § § 6, 9.

90 *Essai analytique*, Preface, vii–viii, ix, x, xvii.

91 Herbert Josephs, *Diderot's Dialogue of Gesture and Language* (Ohio, 1969), 24.

92 *Essai analytique,* § § 524 note, 851, 223.

93 *Ibid.,* § § 173, 223.

94 Bonnet, *Philalethe* (*Oeuvres*, XVIII), Chapter III. This work was written after the *Essai analytique* so it does not predate his distinction there or in the *Essai de psychologie*. More often, and throughout his psychological works, the above distinction between memory and imagination is replaced by another, in which memory alone is assigned the role of preservation and recall, whereas imagination becomes a faculty that enlivens the sensations recalled – sometimes to such a degree that they are mistaken for actual impulsions of objects.

[95] This quote is from Diderot, found, along with the interpretation just presented, in *Diderot's Dialogue of Gesture and Language*, 10–22. This attitude does not surface in Diderot's articles in the *Encyclopédie* (see article on "Encyclopédie") and contradicts that entire project. However, "In almost all of Diderot's writings apart from the *Encyclopédie*, the dominant impression is one of profound skepticism with regard to forms of language that tend rather to obscure than to reveal the distinctions between truth and falsehood." (*Diderot's Dialogue*, 11) His doubts begin as early as the *Lettre sur les aveugles* (1749) and the *Lettre sur les sourds et muets* (1751).

[96] The quotes and interpretation are taken from Hans Aarsleff, "The Tradition of Condillac", 123–7. The translation is mine.

[97] Cf. *Essai de psychologie*, Chapter VIII with *Essai analytique*, § § 207–9, 228. In the latter text, the sentient being is capable of three kinds of abstraction: partial abstraction, when it fixes on a certain part of the object; modal abstraction, when it fixes on the object's odor, color, movement, figure, etc.; universal abstraction, when it considers only what different *concrete* ideas share. All these abstractions are limited to the soul's ability to distinguish some of the particular impressions that make up the concrete idea, and to fix its attention on one in preference to another. "In all these cases, the *abstract* idea is only a *sensible* idea detached by the attention from the whole of which it makes up a part. I thus call all abstractions of this sort sensible abstractions." (§ 209)

[98] *Ibid*., § 229. Cf. § § 225, 226.

[99] Richard Popkin followed his study of seventeenth-century skepticism with an article trying to determine where it disappeared to in the eighteenth century, why Hume was not taken seriously, why the force of skepticism quietly disappeared, not after it was beaten but after it appeared to have triumphed over Descartes' rebuttals. He seemed unable to find any very satisfactory answer. See his "Skepticism in the Enlightenment," *Studies on Voltaire and the 18th Century*, T. Bestermann (ed.), XXVI (Geneva, 1963), 1321–45. I think the answer lies here in the redirection of questions of epistemology into studies of the representative function of signs.

CHAPTER IV

PERSONALITY

In his *Mémoires*, Bonnet divides the *Essai de psychologie* into three momentous subjects: the first – curiously, we might think – is not a specific topic but rather the whole of his theory of psychology, brought together under the "operations of the soul" (Chapters I–LX); the second is habit (Chapters LXI–LXVI); and the third is education (Chapters LXVII–LXXXV). Placing education at the end perhaps needs no explanation. Bonnet says that the preceding psychological principles were not forgotten when he turned to this topic, but were given practical application as a demonstration of their fecundity. Why the section on education should be considered "the most important part of the work", however, is not immediately apparent. Even less clear, taking into account the diversity of considerations lumped together under the heading of the soul's operations – considerations of mechanism, nerve fibers, faculties, language, freedom, etc. – is why habit should be singled out for the status of a special category.[1]

Before pursuing Bonnet's thoughts, and as a means of providing them with some perspective, something should perhaps be said regarding eighteenth-century views on education, and the relation of these views to sensational psychology. For their source, doubtless it is easier to point to Locke than to sum up precisely what Locke taught. How is it, for example, that Helvetius and Rousseau, both of whose educational theories are traceable directly to Locke, find themselves in such fundamental disagreement with one another? Certainly several readings of Locke are possible – one of the reasons, perhaps, that the eighteenth century found him so useful.

To begin with, what, if any, are the connections between Locke's *Thoughts Concerning Education* and his *Essay Concerning Human Understanding*? Some writers, like Peter Gay, would claim without qualification that the connection is "intimate and obvious", that in his educational writings Locke "applied his philosophy specifically to pedagogy", and that his educational program was developed "as part of a total view of the world".[2] Others, like Nina Reicyn, doubt that any connections at all were drawn by Locke when composing his original letters on education to Clarke, and advise against any attempt "to establish a connection between the totality of the *Essay* and the work on pedagogy of Locke".[3] For the rest, much room remains between these two poles, of course, for locating consonances and lines of connection, while at the same time never presuming Locke – whose way was not to work systematically from a general theory – to be constructing an edifice.[4] Though the operations of the mind, so carefully analyzed in the *Essay*, are not mentioned in the *Education*, still one may argue that they are *presumed* throughout: that is the issue here. The more Locke's psychology is seen as essential and preparatory to his educational observations, the more "intimate and obvious" the connections. However, if the reflections on education are viewed as detachable from the psychology itself, one still may fall back onto certain parallels – such as in the function of habit, the association of ideas, the role of language, or in the whole range of epistemological issues associated with representation.

Still, it is the case that the *Education* is connected with virtue, prudence, good manners, and instruction – in that order. Specifically it is virtue – "harder to be got, than a Knowledge of the World", and the first endowment of the gentleman – which is the goal of education.[5] Locke's concern, then, is with proper conduct, and when it comes to instruction, with knowledge that is useful. The *Essay* ends in a chapter often overlooked – or perhaps seldom reached – in which Locke proposes a tripartite division of everything that the mind is capable of knowing. First there is knowledge of "the nature of things, as they are in themselves, their relations, and their manner of operation", which constitutes "natural philosophy", a very broad area since it encompasses, not only considerations of matter, "but spirits also", "whether it be God himself, angels, spirits . . . " (Bonnet's *Palingénésie* itself encompasses and "sketches" in one work this whole field of "natural philosophy", ranging from the latest in natural history to religious proofs, from psychology to speculation on world upheavals and the future lives of humans and animals). Ethics makes up the second branch of knowledge; in this case the goal is not truth, but right, and a conduct suitable to it. It is concerned with "the skill of right applying our own powers and actions, for the attainment of things good and useful." Third is "*the doctrine of signs*", or knowledge of the means by which the other two sorts of knowledge is gained and communicated: "The consideration, then, of *ideas* and *words* as the great instruments of knowledge, makes no despicable part of their contemplation who would take a view of human knowledge in the whole extent of it." Locke then sums up: "This seems to me the first and most general, as well as natural division of the objects of our understanding. For a man can employ his thoughts about nothing, but either, the contemplation of *things* themselves, for the discovery of truth; or about the things in his own power, which are his own *actions*, for the attainment of his own ends; or the *signs* the mind makes use of both in the one and the other, and the right ordering of them, for its clearer information."[6]

Bearing in mind this division of knowledge, it may be seen that Locke, in his *Education*, addressed the second category althought his readers have been determined to link the *Education* to the first and third categories – that is to say, to the concerns of the *Essay*. For this they need not be reproached. If Gay goes further than the texts allow, in assessing a totality at work in Locke's thoughts on education and psychology, if Locke's own interests were different, nevertheless the epistemology was of consequence to educational theory, whether Locke wished to take up that subject or not. Gay's claim that the *Essay* "laid the psychological groundwork for modern educational theory"[7] may be permitted even if it cannot be substantiated in Locke's own educational writings: eighteenth-century writers were quite prepared to give Locke's work all the system it could bear, and Gay's remarks read like a synopsis of the eighteenth-century reading of Locke.

But if the *Essay* laid the groundwork, interpretation diverged over Locke's psychology itself, with two views of his epistemology emerging, and two perspectives on education. In Voltaire's *Philosophical Letters*, Locke was presented to the French in his most sensationalist aspect. Locke himself had reserved a good share of the origin of ideas to the mind's innately *active* powers: from the outset the mind must focus its "attention" for sensory impressions to be recorded, and it must be generally aware ("perception"). With the retention of ideas and the associations made between them, with the separation, comparison, compounding and abstraction of ideas, the mind's activity is all the more evident. In language, which is involved in these latter processes, once again the mind takes an active part. Finally, in the power of acting, choosing or refusing, the activity of the mind is again witnessed. All these active powers make up "reflection", as distinguished from mere "sensation". Locke writes: "But when [the mind] has once got these simple ideas, it is not confined barely to observation, and what offers itself from without; it can, by its own power, put together these ideas it has, and make new and complex ones, which it never received so united."[8]

It is perhaps a tribute to the success of the interpretation of Locke associated with Voltaire, Condillac, Helvetius and others that, even today, Locke's "empiricism" would appear all-too-accessible: we only need mention "*tabula rasa*". Whatever the curtailment of mental activity in Condillac's adaptation of Locke's psychology, memory retains what has been sufficiently impressed upon it. And as for the comparison and joining of ideas, an area where Locke felt "the mind has great power",[9] as well as for the abstracting and generalizing functions of the mind, these are given over to language and incorporated within the doctrine of signs. The tendency in Locke, to make reasoning relatively mechanical, thus increasingly became, for the Lockean associationists the whole of thinking.

In considering the effects of this sensationalist psychology on educational theory, we find that, as Isabel Knight has put it, "An environmentalist orientation toward the nature of man puts a signficant burden on education. If man's mind begins as a *tabula rasa*, anything can be written upon it. Those who are responsible for his rearing and education can make of a man, intellectually and morally, whatever they wish, provided they have the technical capacity."[10] Such is the line of thought that leads directly to Helvetius' *De l'esprit* and *De l'homme*, in which may be found, not the new set of Newtonian laws summing up the world of the mind in a few incisive and simple formulations, as Helvetius had hoped, but rather the last results of the excessive rationalism, reductionism and overly-simple explanations of sensationalist psychology. As presented by Helvetius, they were enough to revolt his fellow philosophes, particularly those most sensitive to the complexity of things (Diderot, Rousseau). "Education is everything", Helvetius concluded, by which he meant that absolutely all diversity in intelligence, character and temperament are to be traced back to sensory experience, to the education of the senses. Every mind being equal (identical) and empty at birth, physical make-up and heredity contribute nothing.

But to return to the point that several readings of Locke are possible, Helvetius no doubt owed much to Condillac and to La Mettrie for his formulations, although he acknowledge an indebtedness only to Locke. Still, such an acknowledgment is not exactly inappropriate. Even the staunchest advocates of the "active-mind" reading of Locke's *Essay* must admit that his choice of the *tabula rasa* image was "unfortunate" for encouraging the notion that the mind is entirely passive, especially when Locke himself sometimes speaks of the mind this way.[11] One finds, in Locke, ambiguities and vacillations in the definitions of significant terms, and overstatements that go so far as to pose contradictions or alternatives. The concept of "reflection" itself embraces too many activities, providing introspective contents, accounting for volition and choice, while also making up the various mental operations directed toward the recollection and association of impressions. Condillac tried to clean this up, but at the cost, perhaps, of merely transferring the ambiguities into language.

And in assessing the power of education, again Locke does not seem to be of one mind. Scattered throughout his *Education* are statements that Helvetius had only to adopt wholesale and without embellishment in order to conclude that everything we are is the result of education: Locke views the child's mind "only as white paper, or wax, to be moulded and fashioned as one pleases"; "I imagine the minds of children as easily turned this or that way as water itself . . . "; " . . . of all the men we meet with, nine parts of ten are what they are, good or evil, useful or not, by their education . . . "; "The difference to be found in the manners and abilities of men is owing more to their education than to anything else."[12] However, this same work expresses opposing views that pay tribute to inherited qualities: "Each man's mind has some peculiarity, as well as his face, that distinguishes him from all others; and there are possibly scarce two children who can be conducted by exactly the same method"; "God hath stamped certain characters upon men's minds, which, like their shapes, may perhaps be a little mended but can hardly be totally altered and transformed into the contrary."[13]

These latter remarks are those recalled from Locke's *Education* when a first statement is found there of educational individualism, leading to Rousseau, Pestalozzi and all progressive educational theory. Each individual is unique, and it cannot be assumed that a more or less uniform process of education will produce uniformly acculturated products. Observation becomes the key to the successful education of each individual; it is so important that Locke and Rousseau argue for the one-to-one relationship of a private tutor. With Rousseau, the teacher studies the child far more scrupulously than the child studies anything at all. The tutor observes, the child does what it likes: that is Rousseau's "negative education". Let it be noted that the tutorial ideal is not at all incidental to a model of education in which the subject becomes the subject-matter, in which surveillance makes up a large share of the content. Correlatively, education becomes deceptive, it employs "psychology", it takes advantage of the differences found between children and adults in the child's underdeveloped reasoning capacities. This clearly is most evident in Locke, who is no partisan of permissiveness, and whose aim it is to teach virtue and to create gentlemen – in every way to direct the child from early on – but who always prefers (for its efficiency) the inculcation of this virtue through games, pleasures, rewards, good example: in sum, through the stratagem of instilling habits unnoticed rather than by means of rules backed up with severe discipline. The tutor is advised to watch for those "favorable Seasons of Aptitude and Inclination", moments when the child is more open and susceptible to influence: "This I think no hard matter for a discreet tutor to do; who has studied his pupil's temper, and will be at little pains to fill his head with suitable ideas."[14]

Helvetius or Condillac consequently find in Locke a mind so malleable that it is simply a matter of filling it with contents. Condillac, showing a certain disregard for the capacities and interests of a seven-year-old, thus proposed to begin his "natural" education of children with a review, during the first month, of all his own principles of the psychology of knowledge. For what reason? That the child, by beginning his education with a clearer idea of *how* he thinks, would forthwith start thinking more clearly. Hence the Locke espoused by Condillac and Helvetius results from a sensationalist reading of the *Essay* combined with the view, occasionally expressed in the *Education*, concerning the pliable nature of the young mind, and, added to this, the the shifting of education's goal from the creation of moral beings to the creation of intellectual ones.

On the other hand, by emphasizing the mind's inherently active powers, its capacity for dealing with received sensations in its own manner, and by joining to these powers the inherent dispositions and talents distinguishing each person, quite a different picture of the individual emerges: an individual actively sifting, weighing, ignoring and interpreting sensations, partly in accordance with his experience and partly in accordance with innate capacities and preferences. What emerges, in fact, is Rousseau's reply to Helvetius in the *New Héloise* and in the *Émile*, a reply to one interpretation of Locke by means of another. The contradictions and obscurities in Locke have been sorted out and realigned into two battle zones. Repeatedly in the *Émile*, and encapsulated in the *Profession of Faith* section, Rousseau reasserts the Lockean division of the mind into its passive and active aspects:

> Then I reflect on the objects of my sensations; and finding in myself the faculty of comparing them, I feel myself endowed with an active force that I did not previously know I had.
>
> To see is to feel; to compare is to judge; judging and feeling are not the same thing. By sensation, objects are presented to me separately, isolated, as they are in nature; by comparison I move them, I transport them so to speak, I put them on top of each other to decide about their difference or similarity and in general about all their relationships . . . . Let any name whatsoever be given to this force in my mind which brings together and compares my sensations; let it be called attention, meditation, reflection, or whatever one wishes; it is still true that it is in me and not in the thing . . . [15]

These passages were originally notes written by Rousseau in the margins of his copy of *De l'esprit*, refuting Helvetius' sensationalist epistemology.[16] Helvetius, he felt, had reduced man to the status of an animal, for man must not be denied freedom of choice and the responsibilities that go with it, and it is just man's capacity for active thought that distinguishes him as a free agent.

From a more empirical standpoint, Rousseau could not accept the sensationalist premise that all minds are the same at birth, that children are born with no special aptitudes or "genius", that every difference in intelligence, character and temperament is the result of education. Education's effects are great, but cannot turn idiots into brilliant thinkers. Nor is it the task of education to mold minds; its task, rather, is to develop the mind's innate potentials, to assist the child, through his stages of development, to come to know his strengths and weaknesses so that he may come to full knowledge of his individuality. True potential is returned entirely to the individual. The educator only assists in a process that remains an interior development, an innate unfolding. In the spirit of his age, Rousseau's considerations of the biological development of the child bear an affinity with preformationism.

Rousseau does not, however, deny the modifying influence of sensations. On the contrary, they pose a threatening interference to proper development. Already in Locke's epistemology – and in opposition to innatist claims that children are born understanding abstract concepts such as God and duty – there was a developmental emphasis with regard to the beginnings of reason. If these considerations did not play a great part in Locke's *Education*, it was because he believed children began reasoning quite early, and certainly prior to the start of formal education. Rousseau lengthens the stages of development, however, making each independent and imbuing each with its own perfection and maturity. True mental *activity*, true capacity for judgment, he actually puts off until the child begins to reason at about the age of twelve or thirteen. Lacking an adult's reasoning powers, children consequently are most vulnerable to sensory ideas about which they have little capacity to reflect. To this essential vulnerability, if we add Rousseau's assumptions that we are born good but become corrupted through the civilizing process, it is not difficult to see how he arrived at his "negative education" as the proper education for childhood. It functions at once to monitor sensible ideas and to provide a proper psychological climate for encouraging the natural development of the child's inner capabilities. With watchful eye, the educator combines scrutiny for latent talents with the embrace of a protective vacuum. Rousseau's denunciation of book-reading before the age of reason is one more part of this general monitoring of sensations, which easily could form habits of taste and of mind before the child's ability to select or reject them reflectively.

But is this not, in effect, a grave concession to the sensationalists? Even though the mind's potentiality always counts for more than the environment, according to Rousseau's genetic psychology children younger than twelve are so susceptible to environmental influences that their environment must be carefully controlled. They are precisely the animals Helvetius made of all mankind. Of course with Rousseau, for whom society poses a corrupting influence, this lack of judgment signals danger and the reason for withdrawing the child: only after the child begins to reason can he start to offset environmentalist influences with ideas of his own, and only then would Rousseau have him carefully instructed in preparation for the corrosive influence of the society he will eventually enter. By extending and absolutizing the age of unreason, Rousseau has extended and absolutized the reach of sensation's power during childhood.

From the foregoing we at least may observe that the eighteenth-century fascination with education had little to do with pedagogy per se, despite the enthusiasm – to a degree perhaps difficult to share – shown for detailed discussions of instructional methods, and despite the spawning of a reformation in pedagogical theory and techniques. Education, rather, must always be viewed in relationship to an epistemology and to a psychology that have come to overlap completely. Descartes' "I think" has retained the self-consciousness it first acquired before the skeptical attacks, now displaced by the question, "How do I think?" It is on a psychology of mental activities that epistemology is built, and psychology is at the same time fully occupied with epistemology, with *ideas*, with their origins, remembrance, expressibility, communicability, associability, abstractability, and so on. As for the demonstration of the theory *in practice* – that is the point of focus upon education. Analysis of education's effects thus wavers between theory and practice, anticipating the responses of the statue, contemplating the domestication of the child, and forever referring responses back to the activity of an appropriate faculty. Hence Rousseau's discussion of the nature of man is to be found in a book on "education", supported by a psychology of human development and of mental activity.

Although we have referred Rousseau's epistemology directly to Locke, it should be noted that his reading of Locke is not unlike Bonnet's. One writer, William Boyd, claims that Rousseau simply adopted the psychology of "his fellow-countryman Bonnet", and employed it against the "prevailing sensationalism." Such influence he does not extend to the field of educational theory, however: "The educational doctrines of the two men have nothing in common."[17] Ironically Bonnet, who had discussed stages of childhood development in his *Essai de psychologie*, had also thought there was an age for the activation of reason, and felt that the yet-unreasoning child should not be subjected to the pretense of reasoning through the memorization and recitation of books which he could not understand; in this Bonnet was of the opinion that here he had anticipated Rousseau with his own thoughts on education following, out of his principles of psychology.[18] Without debating priorities, let us allow that Bonnet and Rousseau share several perspectives, views whose repetition and advocacy in any case indicate their viability.

For Bonnet, who never denies the soul and never underrates the active principle of the mind but whose analysis is concerned primarily with the mechanical aspects of thought, the link between education and psychology is *habit*. Thus we recall his tripartite division of the *Essai de psychologie*: operations of the soul, habit, education. Should this division appear novel to us, it would not have to Bonnet's contemporaries. Locke had not overlooked habit in the retention of ideas; and in the *Education* habit takes on a prominent role. According to Locke, morals and correct conduct are best instilled through habits rather than rules, for they function "more constantly, and with greater facility than reason". "Teach [the child] to get a mastery over his inclinations .... This being obtained, and by constant practice settled into habit, the hardest part of the task is over."[19] The habitual performance is all the more "natural" for its lack of dependence on memory or reflection. This point is useful to the sensationalists, who will argue for a complete equivalence between the "natural" and the habitual. Rousseau denies this, agreeing that "Education is certainly nothing but habit", while arguing, at the same time, that our "nature" is made up of more than just the habits to which we submit.[20] There is no denying that we are subject to the modifications of habit – even to the most unnatural habits, so long as there exists a correspondence between these habits and (social) conditions. But men can shake off these habits, too, reasserting their natures with a change of circumstances. For we have original ("natural") dispositions.

In Bonnet's analysis of habit, he seeks to employ habit in support of his fiber theory of the nerves. "Before me, scarcely any attempt has been made to indicate the first psychological principles of *habit*. Discussion has most often been limited to pure generalities, nearly entirely moral. My manner of philosophizing did not lead directly to generalities of this sort; instead it led me to first seek the foundation of *habit* in the organic mechanism." In summarizing his *Essai de psychologie*, he writes: "I there envisage all the operations of the soul as so many habits, and I see that they are the general effects that must follow from the more or less frequent repetition of the movements in the same fibers. I find in habit the first origin of tastes, penchants, mores, character."[21]

Analyzing the mechanics of habit according to the logic of his fiber theory, Bonnet locates the origin of habit in the repetition of specific movements in specific nerve fibers as they are struck by the impulsion of objects.[22] The tenacity and reinforcement of a habit depends, first, on the frequency of the repetition of the movement, and, secondly, on such physiological considerations as the nature of the nerve fiber (how receptive it is to a particular sensory impulse) and the role of nutrition. The retention of the movement in the fiber constitutes a habit; it is the disposition retained from the action of an object. In general, nerve fibers are most receptive to movements and to the displacement of those movements as they are developing, since growth marks the conversion of fluids into increasingly solid matter, and nerve fibers, originally very soft and malleable, also solidify along with the rest of the body. Nutrition – the assimilation of foreign, "nutritive" matter into the expanding, preformed structure of the germ – increases the solidity of all the organs and parts of the germ as "development" takes place, as the "spaces" of the web-like germ are gradually filled. With growth, then, a habit increasingly becomes embedded in the nerve fiber, because the action of the object that physiologically modified the original state of the virgin fiber, if it is to persist, must continue to influence the arrangement of the assimilated nutritive matter. The more often the movement is repeated, the stronger this disposition becomes, and the more deeply it is embedded. "The fiber grows, its solidity increases, the contracted disposition is strengthened, enrooted, and the fiber daily becomes less susceptible to new impressions."[23] All the particular habitual dispositions, which contribute to the distinguishing characteristics of an individual, stabilize and harden due to the growing resistance of nerve fibers to new determinations and rearrangements: habit, originating in childhood, becomes stronger in youth, firm during adulthood and indestructible in old age.[24]

Mediating or even negating the effects of sensible impressions are the inherited dispositions of the body, "by virtue of which [each individual] is susceptible to specific impressions in varying degrees. Consequently, the same objects do not produce the same effect in every brain. From birth each brain has an accent, a bearing that distinguishes it from every other".[25] Innate structural characteristics are, therefore, every bit as significant as acquired dispositions in determining individuality.

These various relations between the sensory impression and the receptive state of the fiber, leading to the formation of individualizing habits, might be characterized as *passive* syntheses. For such habits are not acquired in acting or in contemplating, but in being predisposed and acted upon. Such passivity can be seen again the destruction or displacement of habits already formed: first, the object responsible for the habit is removed; then, either the elements of the fiber(s) containing the habit are subjected to different determinations, capable of supplanting the former movements, or else determinations, capable of negating or overriding the effects of the original habit, are imprinted on other, associated fibers.[26]

Rational beings are distinguishable from sentient beings in that they are able to overcome pure passivity and to actively affect and even redirect these modifications. Here a distinction may be drawn between passive (sensible) syntheses and active (intellectual) syntheses in the formation of habits. In a sense, all our ideas, even the most abstract, are habits. They all end up as specific determinations of specific fibers. They all end up in the memory, subject either to voluntary recollection or to involuntary imposition. But reflective thought is active: an interaction, on the one hand, of the faculties of thought with the determinations of the object, and on the other hand, an interaction among the reflective faculties that analyze, compare, abstract, and judge. Moreover, interaction takes place between our ideas and already formed habits. The habits of inclination, taste and thought, once acquired (in relation to our particular surroundings and circumstances), have an increasingly marked effect on our perceptions and ideas. While thought and language thus make active intervention possible regarding the formation of habits, it is also true that habitual dispositions extend to the formation and interpretation of ideas, and to the inclinations, desires and passions on which judgments and decisions are made. The relations between our thought and the particular circumstances of our environment, between thought and our already acquired dispositions, and between these syntheses and the preformed structural dispositions of our nerve fibers, thus produce interconnections, negations, neutralizations and interactions so complex that Bonnet's psychology has little difficulty identifying sources of individuality. It may even be possible that preformed differentiation of capabilities applies to the organization of the soul, homologous to the preformation of the body, an idea that Bonnet acknowledges but leaves unresolved.[27]

As for education, which follows directly from Bonnet's analysis of habit, it represents the manipulation, the *formalization*, of habit: under its direction the informal (random) acquisition of habits – of tastes, dispositions, penchants, mores, prejudices, and ideas through chance encounters with one's environment – is superseded, insofar as is possible, by their controlled dispensation. Here the role of education corresponds to that of language. It performs an associative function, imprinting new determinations in the brain[28] and establishing associative connections between these ideas; it performs a distributive function, presenting "a varied series of objects of the senses in a certain order", undertaking, that is, to establish the best, the proper successive order between these imprinted movements; and finally, it "renders the whole thing habitual".[29] Bonnet draws this comparison between pedagogy and language: "Ideas are adjoined (*s'associent*) to their signs as they are adjoined to one another .... This double association of ideas with one another and with their signs constitutes the foundation of the knowledge (*connaissances*) of each individual. The art of teaching thus consists in general in multiplying these associations, in strengthening them and in securing them in an order that assures effects ... ".[30]

Education must not only order our sensible and intellectual ideas but *reorder* them, rechanneling habit formation away from the lower, sensual side of the soul towards its higher, intellectual functions, thus reclaiming rational beings from the "savagery" of the sentient, the unnurtured. For Bonnet, as for Rousseau or Diderot or Locke, education's aim is moral, and its goal is achieved when it "has given a decided superiority to the movements of these [moral] fibers over those appropriate to sensual pleasures".[31] The "moral fibers" are none other than intellectual fibers because all ideas of moral perfection are notions or abstract ideas, which only the understanding is able to produce. Intelligence thus plays a predominant role in all moral activity. The understanding cannot simply generate these notions, however. It creates nothing, acting only on ideas presented; thus it must be placed in the right circumstances to be exposed to these ideas. In this task education holds primary rank. In effect, the art with which education molds the fibers of the understanding decides – or at least plays a large part in deciding – the moral perfection of the individual.[33]

"Education creates nothing, but puts what is created to work." It is the development and actualization of potential; it raises up, out of its natural condition, an "admirable machine" of tremendous capability, but one that nevertheless remains almost pure potentiality. Developing this potential, perfecting what nature on its own cannot – this constitutes the unique power of education.[33]

Yet its force is thwarted in several ways.[34] First, it must root out or displace habits that have already become embedded, a task whose success becomes increasingly unlikely with each repetition of the sensation instituting the habit, as well as with the advance of age. "When education has permitted sensible objects to operate on their appropriate fibers too long and too strongly, it is hardly within the power of a good education to surmount the movements of these fibers by contrary or different movements."[35] The *exclusive* power of education is thus contradicted by Bonnet's entire fiber theory; and it is opposed again by his even more fundamental theory of preexisting germs, which extends preformed ("natural") dispositions to every individual, dispositions that modify or neutralize the effects of particular sensations. Even a cursory reading of Bonnet's educational theory, in the *Essai de psychologie*, shows that he concentrates a good deal of attention upon these natural dispositions. Thus he retreats from the view that would extend absolute power to education, and he is compelled to do so in order to remain consistent with several premises fundamental to his entire system. Talents, temperaments, numerous virtues and vices are included among the natural dispositions. Character formation is the result of an interaction of education with already acquired habits and with preformed dispositions. "In observing the character of my fellow-beings and in meditating on this important subject", Bonnet writes, "I readily discover that characters are as diversified as the circumstances presiding over their formation and their development; and among these circumstances I above all distinguish those of generation and education."[36]

Bonnet follows the view that education must conform itself to the particular aptitudes of the pupil. A wise educator discerns the natural dispositions of the mind (talents) and adapts his instruction to them: the pupil should never be made over into something he is not.[37] Similar treatment extends to the natural dispositions of the heart (virtues). "One is born temperate, humane or courageous just as one is born a musician, a designer or a poet"; but education ennobles these dispositions, "elevating them by degrees to the rank of moral virtues".[38]

Nature can also be "vicious"; the worst dispositions are innate just as are the best. In dealing with these, education explicitly operates on another level, no longer encouraging individuality by assisting the development of natural dispositions, but here working to undermine, modify and redirect natural dispositions. The method becomes covert, artful, insidious: the prudent educator avoids frontal attack, attempting instead to "moderate" thought little by little; he does not want to institute a "fatal lethargy" in displacing the vicious temperament – his aim is not the destruction of the individual – but to contain and rechannel its force.[39] To this end, education is often in contest with the individual's natural dispositions, since they have been judged insufficient or inappropriate. "The force of education modifies the force of *le naturel*. Education is a *second birth* which imprints new determinations in the brain."[40] But according to what standards? How is it known which order of determinations is best? Can it not readily be seen that the aim of education, as the formalized reordering and systematizing of habit, as exclusively an instrument of instruction, is to replace informally acquired habits and undomesticated natural dispositions with the habits of culture, with its ideas, tastes and values?

Yes, education – that domestication of vice, that disentangling and encouragement of talent, that supplanting of the chaotic habits of undirected experience with the manipulated and modulated repetitions of civilization, that awakening first of language and of thought and then of the arts and sciences, that controlled dispensation of reason – it is indeed a "second birth", its limitations unable to conceal its vast domain. It differentiates ("[Education] often renders members of the same family as different from one another as from inhabitants of distant climates"); it develops the individual (it is responsible for the achievements of a Newton); and it develops whole civilizations (it alone has converted the Europeans, once barbaric tribes, into nations of scholars, a feat someday destined to be repeated among the American natives).[41]

Education actualizes not only the potentiality of talent but also the latent power of habits; or rather, the dispersed power of habit becomes concentrated wherever education appears: like language, education's authority consequently derives from its ability to create order, to arrange distributions among the unordered sensations, and to intervene in the unwitting repetitions of these sensations (habits). In brief, by formalizing habit, education becomes a redoubtable constituting force, capable, on the one hand, of negating the effects of acquired (habitual) dispositions, and, on the other, of modifying, redirecting or activating potentiality in natural (preformed) dispositions. So extensive is its power that it is placed on a par with generation: together, they produce that ensemble of natural and acquired dispositions, that "physico-moral ensemble", constituting the individual; and Bonnet designated this ensemble of unique dispositions as "character".[42]

Thus *character* indicates the conjoining of all the particular dispositions, all the differentiating characteristics, that make up a unique individual, that constitute the individuality of the individual.[43] Hence it is this concept of character that encompasses the whole range of our discussion thus far, the object of which has been to focus on the sources and the development *of individuality* within the context of Bonnet's psychology. All the factors constituting individuality are embodied in this concept.

Yet we have been dealing with only one side of the problem, for what gives *unity* to this "ensemble" of habits and predispositions? Are we to conclude – and this certainly is a possibility not overlooked in the eighteenth century – that the individual is no more that an aggregate of accumulated characteristics, whose sum constitutes an *aggregate self*? Dissatisfaction with this answer requires that we become embroiled in all the questions concerning oneness, unity of self, the soul – matters requiring a shift in our attention from differences to identity, from individuality to the persistence of personal identity.

In the thought of Bonnet, the idea of an aggregate self is judged hopelessly inadequate, even self-contradictory, as an explanation of how we, as individuals, may be perceived as integrated wholes, of how all the habits of which we are composed and all the faculties of our thought can be said to embody a unified individual.

> The substance within us that thinks and acts is thus not a composite, like matter, wherein ideas are an aggregate of parts: for, if that in us which thinks were formed of an aggregate of parts, how would it be possible to unite this composite with the simplicity of the sentiment of self (*moi*)? There would exist as many selves as there are parts in the thinking substance, an obvious contradiction of the *sentiment intime* which assures us that, within us, there is only a single self; wherein the unique self would be a result of all the parts, of which the different impressions would become united in this unity that we call *our self*.[44]

Just as "character" sums up the factors contributing to individuality, in Bonnet's psychology, it is with his analysis of "personality" that he takes up the various issues related to self-identity. The concept of personality is concisely delineated in a key passage in the *Essai analytique*:

> Sensations are modifications of the soul: modifications of the soul are the soul itself existing in such or such a manner. The soul has a sentiment of itself; and this sentiment is as inseparable from each of its modifications as these modifications are from the soul itself.
>
> Thus when the soul experiences the impression of an object and when it recalls one or several other impressions at the same time, it *identifies itself* (*s'identifie*) with all of them: and this *identification* is the foundation of *personality*.

Two sorts of personality must be distinguished: the first is that which results simply from the connection that reminiscence puts between antecedent and subsequent sensations, by virtue of which the soul has the sentiment of the changes of state through which it passes.

The second kind of personality is that *reflective* personality which consists in the *retour* of the soul upon itself; by this means separating, in a way, its own sensations from itself, it reflects that it is itself that experiences them or that has experienced them. The being in possession of such a personality calls that in itself that senses, *self* (*moi*): and this self, being embodied, as it were, in all the sensations, appropriates them all and fashions of them a single, self-same existence.[45]

The first two paragraphs spell out the foundation of *all* personality, while the last two describe the two types of personality, sentient and rational.

Animals, children or the statue do not make their own distributions: these are made for them through sensible impressions and through the mechanics of reminiscence. In the case of sentient personality, true, a certain unity of self exists, a successive unity in the self's identification with the order of its impressions as they are received, retained and recognized when recalled in the memory. But this is a limited sort of unity, for the sentient personality lacks any real *idea* of self. It lacks the reflective faculties and the language needed in order to generalize an idea of its self, a self separate from any of its impressions. Unable to reflect that, apart from these impressions, there exists a distinct and unified self experiencing the impressions, it never comes to know its self as itself, for the self is learned of only through reflection. These reflective deficiencies are naturally accompanied by the lack of signs required to represent such ideas: "The statue is still a long way from being able to say 'self,' because it is a long way from being able to reflect on what it senses. . . . In a word, because the statue cannot say 'self' it lacks the idea of self: this idea of necessity requires a representative sign."[46]

Thus deficiency again points to the characteristic impotence of the sentient being: deprived of reflection, and consequently unaware of its self, it lacks the power to impose its own order upon its impressions. Sentient deficiency, as Bonnet makes clear, invariably returns to the incapacity of true appropriation, of truly *re*-ordering the succession of received impressions and thereby bringing them under one's own control. Therein lies all the difference between sentient and rational beings: "The being in possession of such a [reflective] personality calls that within itself that senses, *self*; and this self, being embodied, as it were, in all the sensations, appropriates them all and fashions of them a single, self-same existence." Sentient personality, on the other hand, is coextensive with the sequence of its received impressions: incapable of appropriating impressions, it is appropriated by them.[47]

One thing tempers this bleak portrait of sentient impotence, for, in reminiscence, the beginnings of appropriation and of unity exist: it is not a question of succumbing entirely to the sequence of received impressions, since, by means of reminiscence, these impressions are grouped according to the categories of recognized repetitions (identity) and novel impressions (difference).[48] Just as Bonnet granted sensible abstraction to sentient beings only in the *Essai analytique*, after having first denied them this capacity in the *Essai de psychologie*, he does the same here with memory: for without the functions of memory all that would exist would be the same embodiment, successively attached to and then detached from each isolated sensation, unable to compare its immediate sensation with any previously experienced sensation. "The self would, in a way, be renewed or created anew with each sensation."[49] And in the absence of sensation, what? Nonexistence? A statue?

With this generous investiture of memory, the personality granted now represents the foundation of true personality, and Bonnet's magnanimity, in this regard, is identical in purpose with his extension of sensible abstraction: he wants to replace the concept of sentience as total deprivation with the concept of sentience as *foundational*. Sentient speech (the language of natural signs), sentient thought (sensible abstraction and sensible associations), sentient personality (sentiment of successive continuity in relation to sensible impressions) all now point to sentience as the ground of *rational* speech, thought and personality, as their place of origin. No longer will it be unproductive to look to the sentient when assessing the rational, for there the elements of the latter can be observed and analyzed, like a frozen stage in the emergence of rationality, caught in an endless repetition at the level of what is generative of rationality; it is origin as a perpetual representation of itself as origin, but that never passes "within the true". Sentience in every way signals but in no way achieves rationality. The value of its deficiencies are precisely that they do not suggest a mistake, a wrong turn – sentient personality is not "false" personality – but a perpetual, guaranteed failure of the actualization of the true; sentience represents Other not so much as an insufficient version of the Same – insufficient, but always *revealing*, as though everything had been stripped away but the glimmering origin of the true.

Reserved, then, for the rational being alone is this capacity to turn inward and reflect that it is itself, an entity unified and distinct from all the particular perceptions, which experiences, reflects and acts. Hence the shift from the sequence of perceptions to the unity that precedes them. Hence the *sentiment intime du moi*, a sentiment arrived at only by means of reflection itself, a sentiment of oneness and of self-identity apart from all sensations. Reflective personality penetrates the insufficient unity of the sequence by generalizing it, by appropriating and reordering it; thus true personality dissolves the serial unity of the successive self, recovering unity at the *center*, the ordering center, the center of ordering – the soul. "It is in order to satisfy this sentiment of self, always one, always simple, always unified, that we resort to the existence of that immaterial substance which we call the soul." [50]

Our conviction that our self remains unified and indivisible throughout all sensations, comparisons, reasonings, actions and modifications, such a sentiment of self-unity amidst diversity brings together properties incompatible according to all known laws regarding the actions of material objects, and thus cannot reside together simultaneously in *material* substance.[51] Bonnet elaborates upon this simultaneous double perception of discreteness and unity as evidence of the soul, taking visual images for his examples. One discussion of this appears in the *Essai de psychologie*. When we look at a landscape, we see numerous objects without confusing them. We see them as separate and distinct, but we also see them together as part of a single picture. The question is, how are we able to perceive things both as distinct and as connected, responding at the same time to each particular perception and to the whole without confusing them? It is not a property of matter that, at a common point, it could maintain all these diverse movements, both in their uniqueness and in their relation to an overall unity. Matter would merge these movements into a single composite movement. For Bonnet, only the existence of a special thinking substance can explain it.[52]

Returning to this subject much later, in an essay entitled *Méditations sur l'origine des sensations*, Bonnet analyzes the visual image of a pyramid:

> My soul appears to receive the impression made on the anterior extremity of the optic nerve; but it is in the manner of an immaterial substance: it does not receive a jolt (*choc*) but, rather, experiences a sentiment.
>
> I attempt to analyze this sentiment. The result of my analysis is that this sentiment is one, simple, indivisible: and yet it is that of a very complex object. I clearly distinguish the parts of the pyramid: they are not confused in my soul. I have the *persuasion intime* that my soul apperceives them and distinguishes all of them. I sense intimately that it is in my self that all the partial impressions are united without ever merging. I sense in the clearest way that it is the same self in each partial impression and in the ensemble of all the impressions: that the self that apperceives the point of the pyramid is essentially the same self that apperceives its base; and again, that it is the same self that compares the two impressions and thus judges the height of the pyramid.
>
> This self that is always one in all its operations; this self that itself appropriates them

all or that identifies itself with them all; this self that in the same indivisible instant apperceives, compares, judges, and always has the *sentiment intime* that it is *itself* that apperceives, compares, judges; this self, in a word, that never ceases to be one, identical, although its operations could be infinitely multiplied and diversified; this self, I say, is consequently not *matter*. I see in the organ [brain] only a diverse combination of parts: I find fibers there whose arrangement, interlacing and relations present me with an inexhaustible multitude of particularities . . . . Thus it is not the organ itself that apperceives, compares, judges; for a multiple being would be unable to form this unity, this self of which I have so intimate and so clear a sentiment, and which unites within itself, without confusion, so many things that exist separately outside it . . . . I therefore have the philosophical obligation to acknowledge that my self is not matter.[55]

This is, I believe, a clear statement of Bonnet's position, which requires clarification only on one point. Bonnet often equates self with the unity of the soul, as he has here, but he just as frequently equates it with personality, as when he writes, "I always speak of personality relative to the sentiment each individual has of his self."[54] The simple solution would be to introduce the formula, personality = self = soul, except that the soul is immaterial but personality is not. Personality is connected to memory and memory is contained in the material matter of the brain: the total destruction of memory would necessarily be accompanied by the complete dissolution of personality.[55] Personality must therefore be quite separate from the soul. So how is it that self is, at various times, identified with both personality and the soul? How does it span both substances? It must be admitted that nowhere does Bonnet specifically address this question, and, if clarity and precision do not appear here, then it may well be because the notion of self, and its relation to personality and to the soul, stand too near the heart of his entire endeavor, too near the center of his desire. This only seems confirmed when one considers that the resolution of this uncertain domain of the self forces one to retrace so many ideas of central importance to Bonnet's entire system: the notion of preformation, the mixed nature of our being, and, then, the concerns of the *Palingénésie*, with its speculations on resurrection and on "germs of restitution". But before undertaking a clarification of these latter points, I would like to sketch out the role of the self, in its relation to personality, on the one side, and to the soul, on the other.

The "truth" of reflective personality resides in the nature of self-identity. Although grasped only through reflection, this self is not in itself an abstraction, a product of thought, as has been claimed;[56] rather, it is a *recognition* of an already existing, precognizant unity – that primordial unity of the soul. It is just this recognition that changes everything, for although the ideas stored in the memory continue to make up the content of personality, this content is no longer independent of the soul's self-awareness, of the soul's recognition of its own identity. Previously there existed, on the one hand, increasing content through successive experiences, while, on the other, identity without content or application. The moment (inevitable for rational beings, impossible for all others) of recognition of one's self-identity constitutes the inextricable joining of the soul's unity without content to the memory's content without true unity; whereupon the soul – by means of its various assimilative, reflective and generalizing activities – truly appropriates the contents of memory, turns these contents inside out, and subsumes them under the unity of its own identity. And while the soul appropriates the contents of personality (as memory), for its part, personality acquires the true unity provided by the soul, a unity distinct from the order of sensations. The concept of self thus links personality to the soul; it spans memory and intellect, coupling each to the other, and consequently it applies to both. Hence in "true" personality, in this bonding of personality to the soul through self-recognition, Bonnet silently, without examination, provides the unity of man, divided into two entirely different substances, by bonding the material to the spiritual and producing that double image of the *subject*, that simultaneously empirical and transcendental being.

This subject is both a recognition and an appropriation, an appropriation *through* recognition. Further, to verify that the notion of self applies equally to the empirical and the transcendental, it should be added that this is a double recognition and a double appropriation. As regards recognition, on the empirical side, there is the recognition of recalled sensations in the memory's faculty of reminiscence, whereas, on the transcendental side, there is that recognition of the soul as the center of self-identity. As regards appropriation, on the one hand, there is the appropriation (as acquisition) of all habits, filling in the content of individual character, whereas, on the other hand, there is the appropriation (as assimilation) of these contents to the true unity of self, an appropriation resulting in self-control, in the mastery of sensory impressions through the self-conscious ordering processes of the soul. In both cases it is the first level of recognition and appropriation that defines (and delimits) the sentient personality.

A final point in this regard: character has not been lost sight of, for it makes up the modifications undergone by both substances of our "mixed" natures, modifications recorded in the nerve fibers of the memory, stored there and subject to recall. While the concept of self is concerned with personality from the perspective of unity, of oneness amidst the barrage of impressions and the clatter of mental activities, the concept of character is concerned with the sum of modifications constituting individuality.[57] The former concerns identity (resonance of the same) at the center of personality; the latter concerns difference (richness of diversity), assuring the uniqueness of each personality. Bonnet's analysis of the landscape and the pyramid is thus analogous to his analysis of personality itself – the control and unification but not the merging and dissolution of discrete particulars. Character and self are complementary; together they constitute a conception of rational personality that balances the mutable and the persistent in human nature.

What still remains to be done is suggested by a passage in which Bonnet claims that all his psychological writings, in their concern with personality, serve as a sort of introduction to the *Palingénésie*.[58] Why that work, with all its speculation on the future life of living beings? And why, let us add, do all Bonnet's discussions of personality, wherever they appear in his texts, culminate in speculations on the future? Here, in order to understand Bonnet's analysis of personality within the context of his entire system of thought, we must examine his efforts to bring together his concept of personality and his speculations on a future life. It is a concern that runs through all his psychological writings: thus the *Essai de psychologie* ends in speculations on the preservation of rational personality after death, outlining a view that will be embellished in the *Essai analytique*, in the *Contemplation de la nature*, and, above all, in the *Palingénésie*. The *Palingénésie*, that is to say, is in no sense a work unrelated to Bonnet's other psychological writings. The factors involved in these speculations on a future life were to present, first of all, the most natural (mechanical) explanation of resurrection, an explanation that rejoins his earlier explanation of the origins of life in the theory of preexisting germs; secondly, Bonnet was concerned with providing an explanation of how memory, being material, could be preserved and personality thereby carried over into a future existence.

To begin with this first concern, Bonnet's original speculations on generation and development, found in the first eight chapters of *Corps organisés*, were largely a commentary upon and extension of the views of Leibniz, culled chiefly from several passages in the *Theodicy*. Starting with the idea of the preformed germ, Bonnet argued that organized matter is never actually engendered; generation is simply the development of what preexists in an invisible form. The germ is represented as "the primordial foundation", the elements of which are expanded or "developed" through the assimilation of nutritive particles. As the nutritional molecules fill the interstices, the germ's original organization determines the arrangement of these molecules. Death, in this scheme, is viewed as a sort of "envelopment", a decomposition and separation of the foreign, nutritive matter from the original germ, this germ diminishing to its original proportions only to undergo a second, and far more rapid, development in resurrection. This hypothesis, such as it was, offered a purely mechanical explanation of the dogma of resurrection.[59] But it met with what Bonnet considered to be a salient and irrefutable objection: considering that the germ is stretched through the length and breadth of the body, what is the fate of a being whose limbs have been mutilated? More importantly, what happens in the case of a mutilation of the brain in which the seat of personality has been located?[60] In brief, Bonnet was forced to abandon Leibniz's theory of envelopment:[61] the original preformed germ does not subsist through death.

The setback was only temporary, for Bonnet found another means of establishing continuity between this and the future life: he imagined a tiny organic body to be lodged within the brain of the organized being, a body composed of indestructible matter. This body – yet another germ, a "germ of restitution" – was encased this time not within the reproductive system but within the brain; and this new germ coincided with the "seat of the soul".[62] Thus the seat of the soul, that magical mediator between substances, so often referred to in Bonnet's works within the context of an interface or passageway between body and soul, has here been given a new but analogous role as a temporal interface between the present life and the future rebirth.

Throughout Bonnet's psychological writings, there is recurrent speculation as to the location of the seat of the soul – whether it is to be found in the "callous body", the medulla obligata, or somewhere else. Again, he speculates repeatedly on its material composition: whether it is made up of fire, ether, light, or of some matter analogous to these elements.[63] Such speculations could be recounted in greater detail, but it would be pointless. For the real significance of the seat of the soul is not to be discovered in answers to such questions. One must look, rather, at the manner in which the search is conducted, at the tentativeness of every answer, and yet, a tentativeness backed by the quiet, persistent need to go on speculating. For the seat of the soul is required in principle. Where is it located? In the closest imaginable proximity to the soul, that is all that matters; it is the antechamber of the soul, the double terminus of the nerve fibers and the soul itself, the organ mediating between the material and the spiritual, and through which, from one side, all the impulsions acting upon the various sensory fibers are transmitted and transmuted into modifications of the state of the soul, and inversely, through which the soul acts upon the body.[64] The seat of the soul, that is to say, is to be found precisely in the *center* of the sensory functions of the body; it is viewed as an abridgment of the entire nervous system and as the center from which all the nerves radiate.[65] Only its centrality and its presence are significant in the end. Bonnet recognized this, and in his last discussion of the subject, he suggested that, no matter how far technology advanced, the true seat would probably continue to escape sight and touch.[66] Its elusive presence has, in effect, become a principle: an unwitnessed, imperceptible organ at the center of our physical being. The drift towards metaphysics is clear, where centers are ever out of sight and out of reach. As for the composition of this seat, that is once again unimportant so long as it is able to meet the specific requirement of indestructability; for behind this requirement is the demand placed upon the seat of the soul that, in its additional function as the germ of restitution, it be able to maintain a continuity between our present and future lives. The seat of the soul is "the seat of life",[67] all life present and future, and thus requires a matter that will perdure through death's interval.

How convenient, then, to include among its functions that of the seat of personality, making possible the preservation and continuation of the self through death. Personality, bound forever to physical memory, has all along been in danger of total dissolution at death, but now a solution is found in an imperishable matter, a matter composing the very organ through which all sensations pass, which is centered precisely at the origin and terminus of the nerve fibers and could thus easily be the physical place of memory itself; again, adjacent to, probably adjoined to, the soul, this "organ" would be in an excellent position to remain attached to the soul and survive the disintegration of the rest of the body. Why look further for an explanation of the preservation of personality through death? Bonnet simply argued that the seat of the soul either was the germ of restitution itself or contained "the germ of this new body destined from the origin of things to perfect all the faculties of man in another life".[68] He took the notion of a seat of the soul, adapted it to his notion of preformed, encapsulated germs, and made of this organ a germ destined to develop in the next life. As an imperishable *material* organ, preserving the contents of memory, it fulfilled a fundamental concern for the preservation and continuity of personality, both as delimited by memory and as unified by self-identity.

At the risk of tactlessness, a final question should perhaps be raised regarding Bonnet's considerations of personality. Why the fundamental need for the preservation of memories and self-identity? Here we must acknowledge that, at one level, and no doubt a very basic one, Bonnet's religious beliefs and the personal comfort he derived from them provide an explanation for all his speculations about a future life and for his development of the concept of germs of restitution.[69] In this sense, the preservation of personality is a given, a matter of faith beyond which there can be no explanation and which requires no defense. Given his faith in resurrection, his analysis took the line of the most "natural" explanation possible: he framed his hypothesis with a minimum of miracles.[70] The preformation hypothesis, which he had already supported as an explanation of the origins of this life, was thus extended and modified to explain the origin of the next.

But there is another level of consideration, a level at which the germs of restitution, in tempering or even negating the discontinuity of death, bring us back to the problem of continuity and its nagging reappearance everywhere in Bonnet's system. Here Bonnet explicitly explores various possible conditions of the state of the soul with its attached germ after the death of the body, a period leadingly referred to as an "*intermediate* or *middle* life": it may be comparable to a profound sleep; it may be a state of perpetual dreams; or it may be a state in which the soul leads an active life, continuing to exercise its faculties, or even *developing* them.[71] It is this last suggestion that receives Bonnet's most extensive consideration, and doubtless the one he finds most attractive. Here death represents not even a temporary pause in the life of the individual. A conforting thought; however, and this is the point to be made, it is not a mere fanciful speculation but a requirement of Bonnet's entire system. If personality ends in death, if the identity of the self is severed and memory lost, then the *present* existence of beings is rendered meaningless. That is why these speculations, sprinkled throughout his works, and culminating in the *Palingénésie*, are never peripheral to an understanding of Bonnet's psychology. One must recall that Bonnet's spatial hierarchy of the chain of beings is joined to a temporal progression, which takes the form of an open-ended succession of world revolutions. These revolutions mark out a progression at once physiological (perfection of the body and the senses) and intellectual (perfection of mental capacities).[72] And Bonnet, in his concern with the preservation of personality, in his insistence upon continuity persisting through the catastrophies of death and world revolutions, is primarily concerned with the perfectibility and future perfection of presently existing beings.[73]

For him the meaning of existence has been contained entirely in the ongoing perfection of those beings that exist and are perfectible (that have souls), and it is precisely the recognition of this perfection that gives meaning to our existence. What would be the point of perfection in a future state if it went unrecognized, if former imperfections went unrecalled? "In vain would the ape be elevated to the sphere of man were it unable to retain any sentiment of its first state. *This would no longer be the same being, but another being*. The same would be true for us were death to break every connection between our terrestial state and that glorious state to which we are called."[74] Such an irreducible temporal break signals an irrationality in Bonnet's system just as surely as does a gap in the spatial distribution of beings along the chain. Death signifies a break analogous to that existing, for example, between minerals and plants in the spatial arrangement of beings, and Bonnet undertakes to reduce its magnitude with his chief enforcer of temporal continuity, the preformed germ.

By means of the preformed germ, life has been extended backward, before birth (there being no true engendering), to the original creative act, and forward, beyond death, to an open-ended future (the number of future world revolutions and stages of perfection being left open). But by this very extension has not life been attenuated and demeaned? Death has been suppressed, but at the cost of the virtual elimination of life as a significant category. Life, within a preformed system, has no beginning in birth, no termination in death. We, who live today, represent an intermediate slice, a momentary actualization of a predetermined state within the successive movement from the preformed to the foreordained. Throughout the actualization of this programmed future, though there are physical and intellectual transformations and progressions, though there is increasing *perfection*, is this not development or, dare we say it, evolution, only in a relative sense, from the perspective of the participant? In actuality, from unrecallable beginning to unforeseeable end, every significant variation and modification is predetermined,[75] the playing out of an all-inclusive primordial act.

Preformation precludes any possibility of evolution. Within that system, there has been only one truly temporal moment, a moment, that is, in which time entered in as decisive rather than as simply the numb, successive unfolding of events and beings into their preordained spaces, only one moment in which time was creative, originative; and that moment coincides with the act in which *all* was created, and in which the entire future was ordained.[76] Around us everywhere we see change. We see changes in the growth and aging of things, we see the ravages of time in the geological record; but these are only reverberations radiating out across that vast lacuna of the already created but still empty future, reverberations following upon and gradually fulfilling that one truly creative, temporal event. There is no evolution here, only completion in time's absence; or if not absence – since the temporal dimension is required to unfold the creation – nevertheless, it is as though this creation required time only to deny it. Time merely unfolds, into their awaiting spaces, the continuum of beings and events allocated in advance; and in every way it is the filling of those spaces that is meaningful, a spatial simultaneity constituted through temporal succession: "All the parts of the universe are thus contemporaneous . . . God no longer creates, but preserves."[77]

Hence a final necessity for the retention of personality: for what has been left to free play, to extemporaneous initiation, random circumstance, to intellectual judgment and choice? Personality is the one thing tempering (and thus perhaps legitimating) the preformed, automatic unfolding of "life". With its non-preformed accumulation of individualizing experiences, ideas and dispositions, personality represents the totality of the accidental and the self-constituting, of formation apart from preformation: it is the only index of life as something other than an empty performance, the only indication that we in some way *participate* in our own perfection.

Even here, of course, formation is not unstructured. Both habits of thought and of action are affected by the receptivity of our nerve fibers. But more to the point is the role of education, a topic to which we must make a final return in order to note its tremendous influence and lasting importance. Education, that formidable structuring instrument, never structures aimlessly: all its manipulations are aimed at maximizing the individual's intellectual perfection. It both rearranges and adds to the contents of personality, always with an eye to "improvement", to perfectibility. And since its effects are permanent, since personality does not end in death, the influence of education consequently goes beyond the grave, affecting the level of perfection of the individual for eternity. Our present existence, Bonnet writes, is only the "first fruit" of the germ of restitution. Someday, transplanted to more fertile ground "it will receive the proper cultivation to give its productions all the perfection they are capable of acquiring. Education begins this great work on earth. It prepares the heart and the understanding for the future state."[78] In this life we began as a tabula rasa, gradually acquiring sensible ideas, gradually learning to speak and think. In our future life we will enter with the accumulation of a lifetime of experiences, and the quality of those experiences, attitudes, ideas and capabilities will be of primary importance for ascertaining our level of perfection, our particular link on that expanse of the chain taken up by the human race. There is no other moral retribution in our future state than our relative advance along the chain; and our progression here, as moral and intellectual perfection, hinges largely upon how effectively education has shaped our character.

We, who speak and think, who remember and recognize our interior unity, are free to arrange our own orders. But here education enters, bearing a particular standard of perfection, preoccupied with imparting its particular rearrangements. Consequently, at the upper reaches of species perfection, we find submission once again to an arrangement not our own, this time by choice or by skillful manipulation. This time for our own good. Even the differences initiated by education, in the encouragement of individual talents and tastes, in the multiplication of ideas, only make of education an instrument in the distribution of the continuum in the next world – which is to say that education, by encouraging difference and individuality, never acts as a force of discontinuity, but only as an aid in differentiating the degrees of perfection along the chain of beings. Thus the final demonstration of the power of education. In that one area of our existence that was not preformed, education makes its entry, closing down the possibilities of freeplay, directing and ordering habits according to a culturally inherited model of perfection (sediment of the already formed): it takes up wherever preformation leaves off. The extent to which we are perfected is the extent to which our existence is orchestrated by education. Consequently, where we anticipated freedom we find closure.

For all this time we were moving away from the specter of submission to an order not of our own making, of submission to the order of the sensations themselves. Below the reaches of speech, rational thought, memory and the sentiment of self, even below the animals, there, in the shadows cast by a bend in the chain, resides the Other, this inverted mirror of ourselves: the powerlessness of the speechless, the thoughtless, the unrecalling, the unrecognizing – unmitigated sentience, where sensations are never sorted or appropriated, where the "individual" is endlessly unravelling behind an unremembered, random sequence of impressions. Still, we know this Other, and we know very well why, so many links away, it is of such enduring interest; for it resides within the recesses of ourselves as well, marking our humble sentient origins as well as the terminus of any unretrieved retrogression. It is the visage of disorder and discontinuity, the break that cannot be bridged in the chain (chain of beings, chain of thoughts, chain of events), the gap without reason and without reply. In natural history as in psychology, in this world as in the next, only one thing has meaning: the overcoming of discontinuity – discontinuity ever perceived as spatial gaps remaining to be filled until, in the moment of completion, the lacuna opened up by Creation is replete, and that singular Origin meets its plenary End.

NOTES

[1] Bonnet, *Mémoires*, 170–4.

[2] Peter Gay (ed.), *John Locke on Education*, Classics in Education, No. 20 (New York, 1964), 1, 5.

[3] Nina Reicyn, *La Pédagogie de John Locke: Actualités Scientifiques et Industrielles* (Paris, 1941), 59, 207; translated in John Yolton, *John Locke and Education* (New York, 1971), 9–10.

[4] See *ibid*., Chapters 1 and 2, and pp. 67–8; also M. V. C. Jeffreys, *John Locke*: *Prophet of Common Sense* (London, 1967), 47–9; and Isabel F. Knight, *The Geometric Spirit: The Abbé de Condillac and the French Enlightenment* (New Haven, 1968), 204–6.

[5] James L. Axtell (ed.), *The Educational Writings of John Locke* (Great Britain, 1968), from *Some Thoughts Concerning Education*, § 70. Also see *ibid*., § § 94 and 135, as well as "The Epistle Dedicatory".

[6] Locke, *Essay Concerning Human Understanding*, book IV, Chapter XXI.

[7] Gay, *John Locke on Education*, 5.

[8] Locke, *Essay*, book II, Chapter XII, § 2; see also book II, Chapters IX–XII, XXI, XXXIII, and book III, Chapters II–XI.

[9] *Ibid*., book II, Chapter XII, § 2.

[10] Knight, *The Geometric Spirit*, 201.

[11] Jeffreys, *John Locke*, 55, 57; Yolton, *John Locke and Education*, 52–4, 60–1.

[12] Locke, *Education*, § § 216, 2, 1, 32.

[13] *Ibid*., § § 216, 66.

[14] *Ibid*., § 74.

[15] Translated by Roger D. Masters in his *The Political Philosophy of Rousseau* (Princeton, 1968), 61.

[16] See *ibid*., 62, note 30, and D. W. Smith, *Helvetius: A Study in Persecution* (Oxford, 1965), 172–4.

[17] William Boyd, *From Locke to Montessori* (London, 1917), 44. Boyd in turn apparently took this from Wilhelm Windelband's *A History of Philosophy*, part V, Chapter I, § 7. I am referring to the Harper reprint, vol. II (New York, 1958), 458.

[18] In the section of the *Essai de psychologie* in which he discusses education, Bonnet argues that to educate the young one must begin not only with the most natural and clear ideas, but with ideas that the child is able to understand at his particular age. In this Bonnet carefully follows ideas presented earlier in the *Essai* on the stages of child development, and it is here that he assumed he had anticipated Rousseau. Again, it is due to his demand that education follow the continuity of nature that Bonnet condemns all those works variously entitled "Elements", "Introduction", "Abridgment", and so forth, where the natural chain of ideas and of things is seldom observed. Such works stifle curiosity. (*Ibid*., Chapter LXXXI.) It is better to start with one author, learn his work thoroughly, and them branch out to other authors on the same subject. (*Idées sur l'art d'étudier, Oeuvres*, XVIII.) Bonnet similarly has no use for catechisms in religious instruction, because they attempt to cram everything into a child's head at once in the form of reduced, dogmatic formulas. (*Essai de psychologie*, Chapter LXXXII.) On the relation of Bonnet's method to that of Rousseau, see Savioz, *Charles Bonnet*, 45, 316–18.

[19] Locke, *Education*, § § 200, 64–66, 110.

[20] R. L. Archer (ed.) *Jean Jacques Rousseau: His Educational Theories Selected From Émile, Julie and Other Writings* (Woodbury, New York, 1964), 57. This is from the opening of *Émile*.
[21] Bonnet, *Mémories*, 173.
[22] Bonnet, *Essai analytique*, § 102.
[23] Bonnet, *Essai de psychologie*, Chapter LXII. Also quoted in *Essai analytique*, § 102.
[24] *Ibid.*, § 652; *Essai de psychologie*, Chapter LXI.
[25] *Ibid.*, Chapter LXIV.
[26] *Ibid.*, Chapter LXIII; *Essai analytique*, § 652.
[27] Bonnet, *Contemplation*, part IV, Chapters V, X.
[28] One of the aims of education is to increase the movements in the brain "to the greatest possible extent". (*Essai de psychologie*, Chapter LXVIII.) Compare that statement with Chapter XX of the same work, entitled "On the Nearly Infinite Variety of Movements that *Speech* Imprints in the Brain". (My emphasis.)
[29] *Ibid.*, Chapters LXVII, LXVIII.
[30] Bonnet, *Essai analytique*, § 822.
[31] *Ibid.*, § 516.
[32] Bonnet, *Essai analytique*, § § 513–18.
[33] Bonnet, *Essai de psychologie*. Chapter LXVII. Just how much can be attributed to language? Does it trigger even language and rational thought? Bonnet clearly has this in mind when he writes, in a note added to the *Contemplation*, "Although psychology is unable to demonstrate that man owes his preeminence over the animals to the faculty of speech alone, those born deaf and dumb, and the children found in the woods eradicate all doubt", for those children raised in the forests have been so degraded, "I would almost say so animalized, that they were unable to learn to speak. Those unfortunate individuals of humanity were thus mentally reduced to a level on par with that of the orangoutang; *so it is true that education decides everything in human life*." (Part IV, Chapter VII, note 1. My emphasis.) This note, added between 1781 and 1783, appears to mark a shift in Bonnet's opinion regarding the acquisition of language. For in the *Essai analytique* (1760) he expressed the view that any being endowed, like man, with sufficient power of attention should gradually be able to invent representative signs for himself. (§ 270) The case of children of the woods seems to contradict this view, at least in the short run of an individual lifespan, thus shifting, as Bonnet indicates, more responsibility over to education and acculturation.
[34] Were education in fact able to decide *everything*, why are the children of the woods unable to be taught to speak?
[35] *Ibid.*, § 523.
[36] Bonnet, *Philalethe*, Chapter XIV.
[37] Bonnet, *Essai de psychologie*, Chapters LXXI, LXXVII, LXXVIII.
[38] *Ibid.*, Chapters LXXII, LXXIII.
[39] *Ibid.*, Chapter LXXIV. Another theme, implicit in what has just been said, which Bonnet develops in connection with his Leibnizian principle of continuity, is that education, an art that puts to work what nature has created and does this by following nature, must not modify what is natural in an unnatural manner, by trying to hasten its work and thus leaving things unconnected, by leaving gaps. "A scrupulous imitator of wise nature, [education] does not go by jumps." (*Ibid.*, Chapter LXXV.) Bonnet condemns all pedagogic methods that fail to match the progression of learning with the development

of the body and the faculties of the soul. "It is not within our power to instantaneously double or triple the degree of a talent, to pass immediately from a truth of one type to a truth of another, to discover from the first thrust everything contained in a subject." "The most penetrating and profound geniuses are distinguished from other men only because they have taken more small steps." (*Ibid.*, Chapter LXXX.)

40 *Ibid.*, Chapter LXVII. My emphasis.

41 *Ibid.*, Chapter LXXXIV.

42 *Ibid.*, Chapter LXXXIII; *Philalethe*, Chapter XIV. In the latter reference, climate is placed in the category of a very general modifying cause.

43 Bonnet, *Palingénésie*, part XVI, Chapter I. In the *Essai analytique*, § 652, character is defined as all the sentiments and ideas capable of becoming principles of action.

44 Bonnet, *Rémarques sur quelques passages de l'Entendement pur de M. Kant*, Bibliothèque de Genève, MS 35–5 – 759.

45 Bonnet, *Essai analytique*, § 113. Further on in the same work (§ 200) the second paragraph receives some amplification when Bonnet writes, regarding the condition of the statue: "The soul does not know itself, it knows only through the agency of the senses .... However it does sense what is taking place within itself. It is therefore identical with its perceptions; and we have seen that this identification is the foundation of personality."

46 *Ibid.*, § 114.

47 Elsewhere Bonnet distinguishes two types of growth: growth by "accretion" (simple adding on, as with a stalagmite or a rolling snowball), and growth by "assimilation" or "intussusception". The two types of personality bear a certain analogy to these two kinds of growth, sentient personality relating to reflective personality as growth by accretion relates to growth by intussusception.

48 See the third paragraph of § 113 from *Essai analytique*, quoted above. For more on "reminiscence", the aspect of memory that provides the soul with the "sentiment of recognition" when a recalled idea is recognized (idea as familiar), that provides the soul, again, with the "sentiment of novelty" (idea as unfamiliar), and, finally, that is able, by extension of these two sentiments, to provide the soul with the "sentiment of succession" when it constructs out of the succession of familiar and unfamiliar recognitions a serial continuity of the antecedent and subsequent (total basis of sentient personality), see *ibid.*, §§ 91, 104–7, 193, 318–9, 556, 702–6, 718. In § 718, analyzing the persisting continuity of the sentient personality of the caterpillar as it is transformed into a butterfly, Bonnet writes: "Its self, identifying itself with all the modifications of sensibility and activity, links the present to the past by means of reminiscence; and this connection constitutes its sentiment of its personality."

49 *Ibid.*, § 711.

50 *Ibid.*, § 509.

51 *Ibid.*, "Préface," xxx.

52 Bonnet, *Essai de psychologie*, Chapter XXXV.

53 Bonnet, *Méditations sur l'origine des sensations* in *Oeuvres*, XVIII.

54 Bonnet, *Palingénésie*, part XIV, Chapter II. See also *ibid.*, part XVI, Chapter I; *Essai analytique*, §§ 113–4.

55 *Ibid.*, §§ 706. 711. In *ibid.*, § 708, it is claimed that personality is attached to both body and soul, but "*even more* to the body than to the soul, since memory has its seat in the body". My emphasis.

[56] See Jean A. Perkins, *The Concept of the Self in the French Enlightenment* (Geneva, 1969), 58–9.
[57] Bonnet, *Palingénésie*, part XVI, Chapter I.
[58] Bonnet, *Oeuvres*, XV, "Préface."
[59] See *Essai analytique*, § § 732–4.
[60] Bonnet, *Palingénésie*, part VII, Chapter IV.
[61] See Leibniz, *Theodicy*, § 89.
[62] Bonnet, *Palingénésie*, part I, Chapter I, IV; part VII, Chapter IV, VIII.
[63] On its composition see, e.g., *Essai analytique*, § 738. Bonnet seems to favor a composition of elementary vital fire, as in *ibid.*, § 31, and in *Contemplation*, part IV, Chapter XIII; part V, Chapter XIII. On the use of the concept of elementary fire in the eighteenth century, see Gaston Bachelard, *The Psychoanalysis of Fire* (Boston, 1964), Chapter V.
[64] Bonnet, *Essai analytique*, § 739.
[65] *Ibid.*, § 30.
[66] Bonnet, *Palingénésie*, part XVI, Chapter II.
[67] Bonnet, *Essai analytique*, § 31.
[68] Bonnet, *Palingénésie*, part XVI, Chapter II.
[69] He says as much, acknowledging the comfort his views of the future life gave him, and records, with obvious satisfaction, the comfort they gave to others who read his work. See *Mémoires*, 249–50.
[70] Bonnet, *Essai analytique*, § § 732–4.
[71] *Ibid.*, § 742n.; cf. *Contemplation*, part V, Chapter VII.
[72] See *Essai analytique*, § 745; *Contemplation*, part IV, Chapter XIII; *Palingénésie*, part III, Chapters III, IV.
[73] *Ibid.*, part VII, Chapter II.
[74] Bonnet, *Essai analytique*, § 114. My emphasis.
[75] I say "significant" because this is what Bonnet holds, although he allows nutrition, climate, etc., some influence (of necessity, inessential) in the sphere of physical modification.
[76] Even miracles are preformed in Bonnet's view. See *Principes philosophiques*, part II, 256.
[77] *Palingénésie*, part VI, Chapter II.
[78] *Essai de psychologie*, Chapter LXXXV.

CHAPTER V

THE STRUCTURE OF HARMONY

The indiscreet burgeoning of the *Palingénésie* (from a twenty-five page essay at its conception into several distended volumes) is instructive; and especially so since, from the *Méditations sur l'Univers* forward, this same pressure is witnessed repeatedly, regardless of the subject on which a particular work is focused, to incorporate the entire corpus of ideas from natural history, psychology, physiology, cosmology, etc. True, the weighting of these different subjects, of their constituent ideas and problems, varies; or at least the *space* allotted to each varies according to the presumable nature of the work: now topics are cropped in one area as the price for development in another, now they are expanded again into a verdurous forest of complexities and question marks, all demanding more thought and further research in the future. But never is anything quite let go of, not even for the space of a single text. Never, it seems, is any subject developed without occasioning a silent juggling and jostling for place among the whole vast range of concerns with man and with nature, until the entire array of ideas is marched once more across clean pages under a new title, and in a new arrangement.

Such insistent repetition indicates something other, I think, than a redundant literary style or the absence of an editor: it is a procedure required by Bonnet's relational view of the universe.[1] In a world in which every particular thing is not simply a part of a harmoniously interconnected reality but, more fundamentally, derives its own identity, its locus, from the myriad relations it sustains with everything else, every part implies every other part. Whether it is a problem in natural history such as embryology or entomology, or a problem in philosophy or psychology concerning the nature of human freedom and the activities involved in rational thinking, that problem will find its answer in the web of relations. Naturally no one draws in everything: as human beings we cannot see beyond some very obvious relations, but it remains true that even a partially complete understanding of any particular thing or relation, or group of things and relations, necessarily carries one out into the harmoniously woven tapestry of the whole to look, as far as possible, in the direction of the most obscure and indirect relations, which all have something to reveal, some light to shed, both upon the particular and upon the whole. For knowledge is circular, drawing us out from our starting point into the world of relations that surround and define the particular, and then returning us to where we started in order to see the particular from the perspective of the identities and differences that make up its place within the vast relational field of which it is an element.

Where one begins is not too important: the most piercing gaze would eventually find everything revealed, whatever the beginning point, since everything is connected to everything else. Considered in isolation, each thing can render up only the most partial meaning, and can only be fully understood by grasping the whole as well. In completeness lies truth, in a knowledge of everything. To grasp the simplest relation, it is necessary to fix truly the place of the simplest thing within the tableau and general significance of the whole. Because we are incapable of such full knowledge, however, doubt is unfortunately cast upon everything we do seem to know. Completeness and truth are not meant to be ours. For us, only the sense of their presence, the conviction that a true and harmonious order does exist, and the unfulfilled desire to apprehend it, a desire that drives us, although aware of the impossibility from the outset, to represent that order to ourselves anyway, to somehow recopy it in human terms.

If everything is relational, then human knowledge is relational too. It must stand in a certain relation to the world and to the relations existing between things in the world. And if relations are to be the primary and defining units, this raises certain questions about the relations existing between our minds and the world (connecting knowledge to its objects), as well as questions about the relations existing within our minds (as logical connections between the beliefs, ideas, generalizations, concepts, judgments, etc., constituting the psychological content of knowledge). Since the mind is created of its own substance and subject to its own laws, a question arises as to the connection between ontological relations, on the one hand, and these various epistemological and logical relations on the other.[2] Are the relations between our ideas of things truly representative of the actual relations between those things? In what way, then, and to what extent are we capable of apprehending objects, the relations between them, and, finally, the actual order of reality? These are questions that interested Bonnet, both as a psychologist and as a naturalist; and since his relational view of the world constitutes a primary unity within his texts, it would seem to be worth spending some time on these questions here.

To begin with, the five senses play a role of inestimable importance to epistemological relations since they constitute the bridge between things and ideas. "The body's senses put it in communion with everything around it; through them it is tied to all parts of the universe; through them it in some way appropriates the whole of nature to itself."[3] "In some way" . . . a troublesome three words, however inconspicuously situated. Facing the outside world, nerve fibers, entirely material in themselves, maintain *ontological* relations with the objects whose impulsions they receive and transmit as sense impressions to the soul; facing the soul, however, these same fibers maintain *epistemological* relations with its immaterial substance, since the material impulsions are both transformed into ideas and back again into impulsions[4] (memory being material) to be stored in the fibers and later recalled. The senses, then, are the origin of all our ideas and our sole link with the external realm. What better place to look, in attempting to determine and analyze the formation of ideas, than to the senses, particularly when they, unlike the soul, operate according to material laws, which we are more adept at discovering and understanding? All of Bonnet's endeavors in psychology support this line of reasoning: the secrets of knowing are best sought in the material body, where knowledge originates.

From the perspective of sensationalist psychology, the five senses are important to epistemological considerations not only for the origination and transmission of ideas, but also because, taken together, they make up the boundaries of our entire possible knowledge of the world. Even if it were the case that the relations we establish with reality on the basis of sensations were entirely accurate so far as they went, there would always remain the possibility of distortion through incompleteness. The addition of another sense would certainly increase the range of our relations with the world, but it might also entirely transform our understanding of the world, of ourselves, and of our relation to it. An object acts, relative to our senses, "in direct proportion to the number and quality of the instruments by means of which [we] experience its impressions. Thus there may be beings to whom this world appears very differently than it does to us".[5] There is no need to turn immediately, therefore, to the division between animals and humans, between sentience and rationality, in order to propose different types or levels of knowledge of reality; gradations appear all along the chain in accordance with the quality and quantity of senses. But this also raises questions about the *material* boundaries of human knowledge, i.e., limits determined by the senses themselves. Not too surprisingly, when Bonnet turns to speculations on the future "restitution" of men and animals, the possibility of more sensitive and entirely new senses is foreseen.

How trustworthy are the relations that *do* exist between object and sense fiber on the one side, and between sense impression and mental translation (idea) on the other? Bonnet speaks of all these relations as "natural," by which he means, drawing upon his theories of preformation and specific nerve fibers for support, that they are determinate. Such "natural" relations are most readily demonstrated between object and fiber: that the fibers of one sense receive specific types of impulsions and not others (sounds, smells, tastes, etc.) in itself seems to present a case for fixed relations. "It cannot be doubted that a direct relation exists between the structure of the sensible fibers of each order and the manner of action of the object whose impression they transmit to the soul."[6] These direct relations are broken down even more specifically by Bonnet with his theory that the particular fibers within the same sense are dissimilar, and that specific fibers respond only to certain kinds of odors or tastes. Each fiber thus contains "an original disposition to yield to the impulsion of the object to which it is appointed", and "must be viewed as a tiny machine *destined* to produce a certain movement. This tiny machine's capacity to execute this movement depends originally upon its construction .... The action of the object transforms this capacity into action".[7]

Hence the "natural" relation between object and sense impression[8], for the predetermined is always natural in Bonnet's scheme, recalling that primordial ordering of nature at its inception. But what is "natural", then, in the more difficult translation that occurs between the sense impression and the idea? Here a substance, entirely different and operating according to laws unknown to us, must represent to itself what has been brought before it in the form of material "*ebranlements*". Bonnet finds the most compelling argument for "naturalness" in this case grounded upon the fact that sense impressions impose themselves uninvited upon the soul; and since they arrive independently of any desire or pleasure of the soul, since the soul is helpless to avoid being affected by a specific sensation, the conclusion to be drawn is that sensations are effects, the causes of which, not originating in our will, must derive from something outside of us. That is to say, the soul does not dream up its own ideas; because they cannot be avoided, this again indicates a determinate or "natural" relation existing between sensation and sensible idea.[9] "The first sensations of the soul, its first perceptions of objects are only simple results, absolutely independent of any operation of the mind. They are the primitive laws of our being."[10] The soul's innocence, its powerlessness to refuse sensation, thus becomes a testament to the "naturalness" of the relation between idea and object. What the soul *must* receive is natural, and what it makes of it is natural, too, by once again invoking the assistance of the preordained created order: "The natural relation existing between objects and our ideas is independent of the soul, which is powerless to resist being affected by a certain idea when acted upon by a certain object. The idea is a natural sign of the object, *and this sign is instituted by the Creator*."[11]

By this means of generally expanding his preformation hypothesis into a metaphor, Bonnet determines a necessary connection between idea and object. What he does not and cannot determine, however, is the faithfulness of the representation to the original. This problem leads Bonnet off in a direction unanticipated by his previous argument, on the basis of which we would be ready to hear him pronounce, as a corollary to the "destined" and the "natural", that a *true* relation exists between representation and represented. But epistemological relations are not ontological relations: what brings things together in the real world may be only tangentially related to what connects them in our minds. As will be seen, this presents, for Bonnet, a problem impossible to resolve.

Of necessity, our view of the world is that of our mind. It is the resultant arrangement of the soul's representations, abstractions, and judgments. But these judgments are limited to the information received through intermediaries, the senses, and to the determinate manner in which this information is rendered by the soul. "Thus my soul can judge the qualities of objects only in accordance with the manner in which each sense manifests them. But this manifestation is confined within the rather narrow limits of each sense: consequently, the senses cannot manifest objects, such as they are in themselves, to my soul; they can manifest them only in a determinate relation to the manner in which they act together with that in which the soul apperceives."[12]

Given the circumstances of its isolation as well as the fact of its own peculiar substance and laws, surely it is fair to maintain, as Bonnet does, that the soul cannot be as certain about what takes place outside of it as within it. Yet it is just one of the mysteries of our being that, while the soul must know its own activities best, we, as self-reflective beings, feel ourselves to have been thrust outside those inner workings.[13] True, we have a *sentiment intime* of the existence and unity of the soul, but hardly of its precise mode of operation, knowledge of which we try to gather from the outside, from the effects of the soul's activities on the body. By some odd consequence of the union of body and soul, even in the midst of our thoughts we seem to reside on the side of our bodies, we seem to understand that world – its relations, its laws, its reality – much better, even though *it* is the mediated reality. Surely, as thinking beings, we do in fact dwell within our thoughts, but we *seem* not to: we are rational beings, and yet the space in which those defining mental activities occur are privileged, withheld even from us. We traverse the peripheries of our own center, examining nerve fibers, analyzing "effects," seeking the soul's portal, having dismissed all dreams of entry into the inner chamber. How paradoxical that what we ought to know best, perhaps do know best, nevertheless escapes examination! Trapped within, we feel that we are trapped without.

What Bonnet's analysis of representations reveals to him is that "natural" relations between things and ideas assure us of consistency (the same sensation consistently produces the same idea) but not of certainty (the idea may be inaccurate). Our perspective is found to be characteristically human, the result of a set number of senses and faculties of thought, all with specific capacities and limits, which reconstruct an order that may well have little to do with the actual order of things. The determinate relations according to which the mind represents and acts upon what has been presented to it only attests to the conclusion that knowledge is necessarily perspectivistic: we cannot eject ourselves outside of the relations we are "destined" to maintain with things in order to ascertain the accuracy either of particular relations or of the entire table of representations making up our picture of the natural order. Not that it would be of much use even were we able to gain some distance from the relations locking us into the world, because our perspective would only be replaced by another, equally determinate set of relations, perhaps no more accurate and probably useless for understanding and evaluating our previous perspective: we could only hope for a broader viewpoint, one more sensitive to a larger number of relations. Our very mode of understanding is relational, an interplay between object and soul mediated through the senses. A hiatus thus appears, or at least an uncertainty, in the junction of ontological and epistemological relations, and just as Bonnet had appeared about to pronounce a direct and trustworthy contact with the natural order as the result of the determinate relations connecting us to that order, we find ourselves marooned.

Or do we? For that makes our isolation sound a little too complete. If our relational understanding guarantees that reality can never be known as it actually is, what can be known? Returning to the proposition that the mind does not create what it is helpless to avert being impressed upon it, it can with some confidence be assumed that there is a reality, a real order, and that the relations we establish with it, if not precise representations, are not fabrications either: "I am at least very certain that what [reality] appears to be results essentially from what it is in itself and from what I am in relation to it."[14] Elsewhere Bonnet states this in greater detail: "I will not affirm that the attributes by which matter is known to me are in actuality what they appear to be. It is my soul that apperceives them; thus they have a relation to the manner in which my soul apperceives and cannot be precisely what they appear to me to be. But, assuredly, what they appear to be results necessarily from what they are in themselves and from what I am in relation to them."[15] Knowledge is really the result of an active exchange, an interplay, between the action of the object and the activities of the sense fibers and mental faculties; it is this interplay that produces our relation to reality, our particular perspective. But the interplay is fixed, ordained, preformed: an interplay without free play.

We *would* be entirely adrift were it not for this fixity, uncertain even that a real world existed. As it is, real essences cannot be known, nor can the true ontological relations connecting things, considering that these relations derive from the essential determinations of beings, i.e., from their real essences.[16] In effect, what we do is to reconstruct our own notions of objects out of the diverse properties that we can only ascribe to them on the basis of our received impressions; that is, we represent the "nominal essences" of objects to ourselves. "I consequently admit that I cannot know the real essence of beings at all, and that what little I know of beings amounts to their nominal essence. Therefore, I am well-founded in inferring that it would be possible that a certain property that I judge essential is so only in relation to my very imperfect manner of seeing or conceiving of beings." And, Bonnet continues, in a passage of foremost importance to his understanding of how nature is to be approached, "But this reflection would not prevent me from reasoning on properties that appear essential to me as though they were essential in themselves. It must suffice that they constantly remain the same in relation to me . . . because I acquire the notion of the subject only through the properties that characterize it to my eyes."[17] These words should be remembered when reading through Bonnet's taxonomic endeavors. He is not simply being modest or disarming when he repeats that his idea of the harmonious structure joining all beings in the great chain might have very little in common with the true order, that it is in the end only a nominal order. As for the actual order, we, too, are part of it, positioned at our own echelon, and limited, like all other beings, to seeing the world from our level in the hierarchy. Bonnet always assumes, as well, that beings do exist that are placed higher than humanity. "Beings that possess faculties superior to mine see and conceive of other things that I do not imagine, and their reasonings, like mine, are relative to their mode of seeing and conceiving."[18] In sum: "The order of nature is something very real but it appears under different aspects to the different intelligences that contemplate it. The diversity of these aspects results essentially from the diversity of the relations sustained by the intelligences with nature; *all* these relations are of true realities since they necessarily reault from the nature of the intelligences combined with that of the beings they contemplate."[19]

That is no reason to abstain from contemplating nature. Observation is not thereby rendered useless; in examining our surroundings we still make discoveries – for example, we perceive relations and a relational order – even if our point of view is partial and inadequate.

In observing the beings around me, I soon apperceive that they are not isolated or independent of one another. I discover that they are linked by various relations subordinating them to one another, and that they concur in this way towards a common goal. Again, I find that these relations enchaining the different beings derive essentially from the properties or determinations of the different beings, and, by virtue of these determinations acting upon one another and through one another, that they conspire to produce more or less general effects. I say that these effects are laws of nature; and I define the *laws of nature* as the results of the relations connecting beings. The entire system of these laws thus constitutes the *physical order*.[20]

It would quite naturally be one of Bonnet's concerns to establish that nature *can* be known, however imperfectly, and that knowledge about the world, as well as about ourselves within it, may be increased in breadth and in depth as a result of concerted efforts to understand it; certainly he would not care to persuade the reader that his efforts as a naturalist were completely chimerical. Relations are decidedly the foundational units of Bonnet's study of nature: the very laws of nature derive directly from the relations uniting beings; these beings take on meaning, becoming definable and orderable, to the extent that their relations to one another are uncovered; physiology finds its meaning in the balance and integration of the relations discovered between the various parts and systems of the body.

To what degree these relations can be known depends, first, as has been the focus of the discussion so far, upon the strength of the link between idea and object, upon the authenticity of epistemological relations. Now the mind works at two levels, or forms two types of relations: those linking impressions and objects, and those that are established by the soul on its own through the reflective processes of comparing and abstracting. The proposition put forward concerning simple, sensible thoughts – that they are really thoughtless, automatic representations – cannot be directly extended to cover this second group of relations. But Bonnet does try to maintain the same line of argument, holding that even the most abstract ideas, and the relations discovered between them, have their origin in the things themselves and in the relations sustained in nature. This line is pursued, of course, in anticipation of the challenge that as our thoughts become increasingly abstract (the more we reason) their relation to reality becomes increasingly arbitrary. In essence, Bonnet simply insists that all our abstract ideas, as well as all the purely logical relations that we discover, have their foundation in nature and therefore are not creations of the mind.

It is often repeated in Bonnet's psychological texts that the mind does not create ideas; it creates nothing, only acting upon what has been created (what it receives as impressions from created beings). Analogously, it is Bonnet's position, concerning embryology, that preformation precludes true generation, i.e., that so-called "generation" is only the unfolding of what was created long ago and is misperceived by the layman as an act of creation: with this two-pronged attack directed at the origin of living beings and the origin of ideas, both aspects of which are grounded in preformationism, all creation beyond the original act of creation has been effecitvely exorcised. Relations are no more created than are ideas[21] : logical ideas and relations, Bonnet tries to show throughout his analysis of language and reflective thought, are only generalized sensible ideas that have not lost their connection to their origin in the process. Hence pure imagination is suppressed as the price for preventing *all* ideas and relations from becoming empty fancies; acts of creation turn out to be acts of discovery; inventiveness borders on loss of contact with reality.[22]

It is always the relations between things themselves that account for the relations between ideas, just as it is always properly the case that "my ideas are the representations my understanding has formed of things, the original impressions of which were transmitted to it by the senses".[23] The more we study nature and its relational order, then, the more accurately we should be able to represent that order. Taking this to its extreme, Bonnet can conclude that "Every theory is only the chain of natural results that reflection caused to be deduced from experiment and observation".[24] In general, however, his position is more cautious, thoughtful and mitigated. This brings us to the second issue regarding the knowledge of relations. Assuming that something can be known, the question is how to proceed.

In studying ontological relations, it becomes clear that they are not all equally accessible. Some are perceived as so simple and direct that they are appropriated at first glance, as if by intuition; others can be perceived only indirectly by first traversing a succession of intermediate ideas and thus must be worked towards progressively, proceeding from the most to the least direct relations. "I soon recognize", Bonnet writes, "that *all my knowledge* is, in the last resort, reduced to knowing which immediate or intermediate relations connect things to one another and knowing the order in which these relations coexist with or succeed one another".[25] It is a statement whose importance for Bonnet's epistemological considerations can hardly be exaggerated, applying equally to his taxonomy, and his studies in physiology and in psychology.

Direct relations, presented so immediately to the understanding, according to Bonnet, that there is no intervening operation of appreception, constitute "evidence"; i.e., the connections they present assure the greatest certainty, and they are to be considered "axioms" or "primary truths".[26] The understanding (or "intelligence"[27]) is so constituted that it cannot but surrender to evidence the moment it is presented: "I want to say that intelligence apperceives the relations that are within its scope just as the faculty of sensing apperceives the sensible qualities of bodies. It is no more within the power of intelligence to not apperceive a certain relation than it is within the power of sensibility to not be affected by heat at the presence of a warm body. Therefore, when the understanding obviously apperceives the relation existing between several ideas, it apperceives a *truth*. It acquires this truth instantly, and its acquisition is the affirmation of this truth."[28] That we cannot but acknowledge evidence at the instant it is presented in itself demonstrates "that truth is the object of my understanding".[29]

But often the sense of perfect certainty, of the irresistible force of a direct relation, is not there, and the relation must be sought with the help of our rational powers. In the case of indirect relations, beginning with what is best known, traits of resemblance or analogy are sought with the relations we are trying to find. By way of comparisons we move through intermediate ideas that "are so many links of the chain joining the two ideas whose relation is sought by the understanding. The number of links, the length of the chain, varies according to the indirectness of the relation".[30] Unfamiliarity with the things being compared, the complexity, subtlety, or concealment of the relation, or simply the distance between things with respect to the scope of our understanding, are all factors that can constitute the indirectness of a particular relation: not at a glance does one apperceive, for instance, the relation of the world's existence to that of God's, or the relation of the ideas of truth and beauty to other ideas.[31]

Indirect relations must be ferreted out, something that can be accomplished only with the help of the intelligence: not an intelligence spinning daydreams, however, but one acting upon received information. Observation and experimentation remain the foundation of all methodology: "Since the ideas that my understanding forms of things and of their relations are species of representations of those things, it follows that the representations will be proportionally more faithful and truer as they express the nature of those things and their relations with greater precision."[32] The attention must therefore be applied with great regularity to the things under examination, a well-trained attention in itself being "the spirit of observation".[33] With its help we fix our gaze on objects and become conscious of traits that had earlier escaped us; again, with its help, we are able to hold several ideas in mind at the same time for comparison. Such comparisons are critical, because from these comparisons judgments follow, and judgments constitute the actual revelation of relations.[34] These judgments are themselves ideas that may be subjected to further comparisons and judgments, each judgment in this procession delivering an increasingly general idea. Such a generalizing capacity, when properly grounded in observation, enhances the discovery of relations by bringing into proximity things and ideas that at first sight would have appeared far apart. For us, the order of things is consequently paralleled by an order of ideas by which we attempt to appropriate the primary order through representation and denomination. Our ability to acquire and compare ideas constitutes the very nature of human understanding, according to Bonnet, and the number of things it is able to compare, the number of intermediate ideas it can employ, determines the level of indirectness of the relations it can know.[35] Thus observational data must not only be gathered but assembled, compared, arranged, abstracted: only then can the less immediate relations be seen and the laws of these relations discovered.

More generally, Bonnet looked for a regulative method or logic that would be responsive both to the processes of abstract thinking (chain of ideas) and to the relational structure of reality (chain of beings) activating those processes. He spoke, concerning this, of a "living logic," one that would be "truly practical"; however, such a "universal instrument for the mind . . . will only be instructed by a philosophy in possession not only of all the aspects of rational philosophy, but one that moreover will be master of the most interesting aspects of physics and natural history, and that will also reflect deeply on the nature of the human mind, on the subordination of its faculties and on the manner in which they develop from one another".[36] A logic, that is to say, that would encompass both natural history and psychology, both ontological and epistemological relations, binding them together by its own scope and procedure – rather too wishful a resolution perhaps to the problem of solidifying the relation between epistemology and ontology, but nevertheless sketching a connecting unity between Bonnet's diverse enterprises, and expressing his ongoing dissatisfaction with contemporary logics. Those logics "correspond only imperfectly with the goal of science: they are too confined; they abound in precepts more than in examples, and in varied examples; they are unable to deal with action; finally, they sharpen the subtlety of the mind more than they increase its forces".[37] Magnificent abstractions are produced, airy relations drawn, and all in disregard of the actual properties of and relations among beings. Reasoning, by itself, often deceives. That is why experience must be consulted and why a "truly practical" logic is needed. A chain of reasonings, to be useful, must always be initiated by ontological relations themselves.

How is proper reasoning initiated naturally? In the *Essai analytique* Bonnet finds his answer in a psychological analysis of "surprise". Surprise is obviously brought about by the unexpected: a meteor appearing unannounced, for example, would be a surprise.[38] Just the opposite of habit, surprise arises from the discontinuity revealed in a comparison made within the soul between past continuity and a sudden modification. Its effects can be devastating. "Surprise can reach the point of severely agitating the entire mechanism." So great can our astonishment be at an observed lack of relation "between my ideas and the sudden appearance of the phenomenon", at the appearance of a discontinuity in the connected fabric of our thoughts, that the jolt can result in shock, breakdown, freezing of the faculties. Of necessity, then, are we creatures of habit, for we are creatures of continuity. Such radical results lie at the extreme, of course: so wide a gap opens that we fall dumbfounded into the dread abyss of incomparable, irreconcilable Difference. Less "deranged" sequences, less agitated tremors, lead to entirely different results, producing such pleasing experiences as the sentiment of "novelty", and such beneficial effects as the introduction of observation and method. Mild "derangement" excites the interest of the attention; our thoughts become fixed on this apparently unrelated phenomenon, and with a sense of pleasure (the pleasure of novelty, and the pleasure of puzzles) we examine, compare and reach a judgment. Hence the true beginnings of right reasoning as natural reactions to the (apparently) discontinuous.

To extend such natural inclinations into a comprehensive logic would be to achieve Bonnet's "practical logic". "This art", he writes, "by which I succeed in refilling the gaps (*vides*) separating two or more things to my eyes; this art by means of which I succeed in discovering the relations connecting things to each other, I call the *art of reasoning*".[39] Correct reasoning, whether the stimulus be pleasure or anxiety, is thus induced by the urge to reestablish continuity, and by a psychological uneasiness with discontinuity. Bonnet never doubts the validity of this urge, never doubts that fundamental relations raise everything out of isolation and place all within the continuum of reality. It is made the foundation of proper method. To doubt the existence of such a continuity would, in fact, be unthinkable for Bonnet: it would send reverberations all across the distance of the taxonomic chain that he denominates, and upwards as well to the highest reaches of the chain of ideas, clear up to the most abstract. No, in the continuous space of this world everything is not only related, but related to *everything* else. Thus there exists a harmonious whole. Discontinuity resides only in the lapses of our thoughts, signaling in every case an error in reasoning, the locus of ignorance. Our goal could not be clearer: to surmount ignorance and to discover as many relations as we are able.

No amount of effort, to be sure, can give our search for indirect relations the certainty that direct relations provide. But with careful observation, with the scrutinization, dissection and reassembly of details (analysis, synthesis), and with a logic fully cognizant of the relations holding between ideas and things ("practical" logic), the degree of uncertainty can be reduced, leading Bonnet to speak, in this connection, of probabilities: "Considering certainty as a whole, I can divide this whole into parts or degrees that will be parts of degrees of certainty. These ideal divisions of certainty I will call *probabilities*. Consequently, the degree of certainty will be known when I have succeeded in establishing the relation of the part to the whole."[40]

Bonnet does not simply treat probabilities as statistical percentages, however; they, too, are incorporated within the relational order since the mind acquiesces to probability as necessarily as it does to certainty. It "cannot apperceive the probability of a thing without affirming the probability of this thing"; even here it "apperceives things as they reveal themselves to it or in accordance with the relations it sustains with them".[41] But probability does not have the same effect on the soul as does evidence. The lack of full certainty that necessarily accompanies probability leaves the understanding with a desire for More, a longing never experienced when evidence is presented.[42]

A qualitatively different psychological feeling accompanies the fullness of evidence in direct relations on the one hand, and the incompleteness of probability in indirect relations on the other; analyzed at this level, probability and certainty are juxtaposed according to the difference in the effects produced upon the understanding at the reception of each. At a more profound level, the "truth" of evidence is only human truth, satisfying our particular sense of completeness. Our psychological reactions may determine what constitutes evidence to us, but *both* certainty and probability are essentially uncertain. Our distinct reactions only affirm our perspective: the relations that we apperceive as the most immediate may, in reality, be distant or partial or erroneous, deriving their sense of immediacy from our particular sense or mode of understanding; and these same considerations make our perceptions of indirect relations even more dubious. "The understanding can therefore mistakenly regard something very uncertain as very probable."[43] "Consequently, I conceive that things which to me appear separated by great intervals may approach or even appear to be touching one another in the eyes of intelligences superior to mine. Again, I conceive that if my senses were perfected or if I acquired new senses, my understanding, since it is essentially limited in its exercise by the scope and the number of my senses, would be perfected in the same relation and it would know a multitude of things and relations that completely escape it in its present state."[44]

Not only the gaps that surprise us but perhaps also the relations we find to refill them reside in the nature of human understanding rather than in nature itself. An essential vacillation thus seems to exist in Bonnet's texts. On the one hand, there is Bonnet's perspectivism that still permits knowledge of reality, knowledge that can be increased through diligent observation and proper reasoning. And, on the other hand, we are faced with a rather profound ignorance of the connection between the relations in our thought and those in reality. The latter position is generally submerged, its implications never elaborated at length, and is generally overridden by the sheer fact of Bonnet's numerous observations of and reflections upon nature. But it is there nonetheless and presents an important tension within his works. It is interesting that this position appears at all, for it presents Bonnet with a philosophical view that threatens to cancel his work as a naturalist and even as a psychologist. The questions raised for epistemology by this uncertainty regarding a fundamental link between the order of things and the order of thoughts occupy a place strictly analogous to those questions raised, in the consideration of language, concerning the link between the natural and the artificial sign.

What knowledge we have of the physical realm, Bonnet finally concludes, is all analogical: things demonstrating the same characteristics are assumed to contain the same properties. We do not, for instance, test every new fire in order to assure ourselves that fire is hot. The key proposition of analogy is "that precisely similar effects presuppose the same causes".[45] But this is never more than an assumption, since it is quite possible to conceive of the same effects resulting from quite different causes. Our problem remains that we know nothing of real essences. "The more I mutiply my experiences of the same type of things the greater probability my conclusions acquire. Perfect certainty here is contained in the knowledge of the *totality* of these things, but my experiences are incapable of embracing this totality: still, the greater the number of particulars my experiences embrace, the greater the probability."[46]

The truth of the matter would seem to be that the physical order cannot be demonstrated, and Bonnet can be found admitting that our most profound reasoning convinces us that our natural faculties are unable to demonstrate even the existence of bodies.[47] Yet, characteristically for him, the most striking result of his analysis of our strictly analogical access to the physical order is the "very practical principle" that although our knowledge is only probable at best, we are forced to *act* as though we knew the world with complete certainty. If, in practice, we took such reasoning as our guide we would perish without learning even to drink or eat. "Nothing in fact is more rigorously demonstrated than this necessity imposed upon me by my physical condition; since if I refused to submit to it, I would be a most wretched being, and would even be unable to preserve myself."[48]

For us, perhaps the most striking result of his analysis is also the reason for the swelling of his texts. "Perfect certainty is contained in the knowledge of the totality . . . ". It may be humanly impossible to embrace this totality and to follow up all those indirect relations to their end, but thoroughness requires our attempts at such a totalization through the marshaling of ideas from every sphere of thought. Such a project for a "total" understanding *of nature* is comparable, allowing for the fact that time had replaced space as the dimension in which totalities were to be established, to those efforts since the nineteenth century to write total *histories*, i.e., total descriptions that would draw all phenomena around a central meaning, spirit or world-view, thus reconstituting the form of a civilization by discovering a common significance and cohesion that would subsume all the phenomena of a period. Such histories, too, presuppose a total system of homogeneous relations.[49]

What should be remembered in any case is that, in the harmonious and unified structure of nature, relations form the fundamental units. And for observers of nature they are the object of all inquiries. Bonnet never doubts the reality of their significance in the determination of the structure of the whole or the meaning of the parts. As for the configuration of these relations – the image of the chain or scale, for example – speculations put forward in this regard are not meant to be taken too seriously. After all, this is already something that goes far beyond our limited capacities of perception. The repeated use of the idea of a chain, in particular, gives this notion greater visibility in Bonnet's texts than his relational theory; yet, while he goes very far in trying to demonstrate a relational reality and in trying to discover what its implications are for epistemology, it is not likely that he would have defended, or even felt the need to defend, the idea of a chain of beings at any great length. To regard the chain as a hypothesis is to misunderstand its value, which is really figurative: it is presented as a means of envisioning the physical order and is not proposed as an actual representation of it. If Bonnet employed the configuration of a chain most often and most fully, that is because it was most suitable to the representative sweep of his taxonomy, incorporating in a single metaphor the essential aspects of distribution, connectedness and hierarchy. Certainly it does not exhaust the imaginable configurations of relations, nor is it the only one Bonnet made use of.

I find Bonnet playing recurrently with two other configurations: concentricity and networks. The former appears with his use of the term "systems"; the movement outward from the most "particular" systems (generally a reference to individual living beings) to the most "general" system (the entire creation) is as though in a series of concentric circles. The "sphere" of each being's system "is itself contained within another sphere: the latter within still another; and, the circumstances continually expanding, this astonishing progression is raised by degrees from the infinitely small to the infinitely large, from the sphere of the atom to that of the sun, from the sphere of the polyp to that of the cherub. There exists, therefore, a mutual correspondence between all the parts of the universe; no part is isolated".[50] These concentric systems are themselves envisaged as "enchained" to the broader systems to which they are joined[51]; consequently, they do not present an alternative to the chain so much as a complementary means of imagining the same relational order.[52]

In the *Contemplation*, referring to the naturalist Donati's *Essai sur l'Histoire de la Mer Adriatique*, Bonnet writes: "He did not envisage the progression of beings under the image of a chain: he believed, instead, that they must be envisaged under the image of a web, all the threads of which are joined to one another."[53] Bonnet only mentions this arrangement, offering no criticism of the idea, but it is readily seen that such an idea, applied to taxonomy, would do away with the single hierarchic scale in contrast to all Bonnet's efforts to establish an external distribution of beings according to their increasing "perfections". But Bonnet, too, employs the idea of a network whenever he turns to the internal space of beings. What are preformed germs if not collapsed webs? The principle of germ expansion, of "growth", amounts to little more than the insertion of foreign matter into the interstices of the web. Again, the bond between all the organs of a living being, the fact that each presupposes every other, derives its validity from this image of an organically structured network. Webs provide a more suitable image of relational ties here than do chains because, when turning from external identities and differences to internal organization, rank loses its privileged status in the dissemination of order. The body functions as a unity; every organ is important; it is the contribution of each to the whole that must be grasped for an understanding of the physiology of beings, not the privileged status of any particular organ or function. Finally, for psychology, the image of a network is applied to the physiology of nerve fibers in order to explain the relations or communications between them, without which no association of ideas would be possible.

The relational structure of reality and of our thought is never doubted although the actual configuration of relations remains uncertain, as indeed it must, since to grasp the true configuration would be to know the physical order itself. We only attempt to represent those relations as best we can, to imagine that order as our capacities allow.[54] It is still science, for all the sciences remain, strictly speaking, human sciences, rebounding back upon our own perspective, limning nothing so clearly, perhaps, as our own relation to the relational order. But this should not trouble us: we strain to perceive relations, we project configurations, we try to represent the order of beings in our taxonomies – because we must; because, with the advantages of speech and reflective thought, we have been granted a status denied the other beings we see around us: that of having the capacity to order, even if incorrectly, the capacity to appreciate regularity, harmony and continuity, the capacity to be aware of the existence of order and thus to seek the Order. Failure is assured. It is not within our capabilities to achieve certainty; that misunderstands our place and the limited scope of our knowledge and our science.[55] What we do know, however, is that we are confronted with an immensely complex and interrelated whole, each part of which bears some relation to and sheds some light upon every order.

That is why Bonnet finds none of his projects unrelated, and why a thorough analysis of any one of them must eventually include them all. Perhaps it is the case that, for those with a sufficiently penetrating gaze, for those supra-human intelligences bounding the upper side of the chain, every outward glance, in any direction, eventually wends its way back to the point of departure: that in tracking the most distant relation one recovers at the end – the origin, filled with the meaning of the whole.

NOTES

[1] Bertell Ollman's important work, *Alienation: Marx's Conception of Man in Capitalist Soceity* (Cambridge, 1971), demonstrates that much criticism of Marx has been due to an inadequate understanding of his relational approach to reality, an approach directing Marx's whole use of language and concepts just as it does in Bonnet's case. See the first three chapters of that work, and Appendix II added to the second edition (1976), for much useful information regarding the implications of a relational philosophy. By Ollman's account, it decisively affects Marx's interpretative method. That its displacement was abrupt and virtually total in the course of the nineteenth century is testified to in the misrepresentations of Marx even by early Marxists, and further, in the difficulty of its recovery; for what now appears to be a basic tenet of Marx's thought had for the most part escaped acute interpreters of Marx before Ollman. I think this chapter on Bonnet should lend substance to the view that relational philosophy was not uncommon from Spinoza and Leibniz through Hegel. Bonnet's own assumptions in this respect were likely influenced by Leibniz. In describing, for instance, the relational ties between the monad and the world, Leibniz writes, "there is no term so absolute or so detached that it doesn't enclose relations and the perfect analysis of which doesn't lead to other things and even to everything else, so that one could say that relative terms mark expressly the configuration which they contain." *Nouveaux essais sur l'entendement humain*, quoted in *Alienation*, 32. And in the *Monadology* one will find such statements as the following: "Now this *connection*, or this adaptation, of all created things to each and of each to all, brings it about that each simple substance has relations which express all the others, and that, consequently, it is a perpetual living mirror of the universe." Philip P. Wiener, ed., *Leibniz Selections* (New York, 1951), 544. For an interpretation of Leibniz's position on relational properties and an analysis of his attempt to formalize logical relations, see Hide Ishiguro, *Leibniz's Philosophy of Logic and Language* (New York, 1972), Chapters V and VI. Condillac also believed the universe to be a vast interconnected system of relations. See Isabel F. Knight, *The Geometric Spirit* (New Haven, 1968), 60–1, 130. Differences between Bonnet's considerations and those of Leibniz certainly arise over the question of how reality and thought are related – a central interest of this chapter. Another point of interest: as a consequence of focusing on Bonnet's relationalism – something that other interpreters of Bonnet have passed over in silence – the idea so often associated with Bonnet as central to his view of nature, that of a chain of beings, is displaced and given its proper position, I think, relative to Bonnet's thought. For that, see the last pages of this chapter.

[2] I am indebted to D. W. Gotshalk's *Structure and Reality: A Study of First Principles* (New York, 1937) for its discussion of relations, in the course of which these different categories of relations are distinguished (97–107). Bonnet refers to all relations simply as relations, although he uses them in the senses understood here when analyzing the problems of thought and reality. The categories make the problems clearer.

[3] Bonnet, *Analysée abrégée de l'essai analytique* in *Oeuvres*, XV, art. IV, 8.

[4] "*Ebranlements*" was the standard term used, and not just by Bonnet, to signify the material agitations produced in the nerve fibers and to represent the action of the object or the idea of the soul in the form of a particular mechanical motion. But he also uses other terms to identify these motions, and I have referred to them here as impulsions or impressions. See Isabel F. Knight, *The Geometric Spirit*, 119.

[5] Bonnet, *Essai analytique*, § 199.
[6] *Ibid.*, § 604, Cf. above, Chapter III, note 53.
[7] *Ibid.*, § § 610, 616. My emphasis.
[8] See *Ibid.*, § 218.
[9] It is possible, Bonnet admits, that the entire "course of nature" takes place inside of us. But even if true, nothing is changed, since we are still forced to recognize the same variety, harmony and order of coexistent and successive ideas – ideas by which we continue to be affected in the same involuntary manner. Appearances are precisely the same as in the assumption of a real universe. Intrigued by Berkeley's idealism, Bonnet nevertheless found the identity of its results with the more common assumption of an existing, physical universe a major shortcoming: it neither solved nor changed the nature of any of the problems. If it was Hume's position that Berkeley's arguments neither admitted the slightest refutation nor caused the slightest conviction, Bonnet's view apparently is that they do not admit the slightest refutation nor does it make the slightest difference. See *Philalethe* in *Oeuvres*, XVIII, Chapter XI.
[10] Bonnet, *Essai analytique*, § 201.
[11] *Ibid.*, § 218. My emphasis.
[12] Bonnet, *Philalethe*, Chapter IX.
[13] That one is, in general, more readily convinced of the existence of the body than of the soul is apparently due, at least in part, according to Bonnet, to the mind's being so constructed that it finds it easier to consider its exterior than its interior. *Essai analytique,* § 2; Preface, xiv.
[14] Bonnet, *Philalethe*, Chapter IX.
[15] Bonnet, *Essai analytique*, Preface, xiii. In *ibid.*, § 238, he writes, "The determinations of substance are thus the *relations* under which we apperceive it. These relations are the result of its activity combined with ours." Cf. also, *Philalethe*, Chapters XVI, XV.
[16] *Ibid.*, Chapter VI; *Essai analytique*, § § 40, 856–7.
[17] Bonnet, *Philalethe*, Chapter XI.
[18] *Ibid.*, Chapter XVI.
[19] *Ibid*. My emphasis.
[20] *Ibid.*, Chapter XI. The definition here of laws, as resulting from relations, has its own history in Bonnet's thought, which can be traced back to the *Esprit des Lois*, with which he had some familiarity even before its publication. Mussard, *conseiller et secrétaire d'État de la Republique* in Geneva and Montesquieu's friend, brought the manuscript to Geneva to be published there in 1748. He read several chapters from the manuscript at a gathering attended by Bonnet, who was greatly impressed, in particular with the "Préface's" discussion of laws. Yet he felt that the definition of laws ("Laws are the necessary relation that derive from the nature of things") was in error: laws are not themselves relations, but ought to be understood, rather, as "consequences of relations". Bonnet expressed this opinion in a letter to Montesquieu (April 1, 1754), and added the following example: "The structure of the magnet is the basis of its relations to iron; attraction is an effect, a consequence of those relations." Not persuaded, Montesquieu replied that "The laws of the universality of beings are consequences of nothing, but produce endless consequences." Bonnet did not pursue "this small controversy" with Montesquieu, who died in February of the following year, but he remained convinced that Montesquieu was mistaken. Laws do not derive from the "universality of beings". "Only individual beings exist, linked to one another by a thousand diverse relations .... But the relations

that link beings are not actually laws. They are the foundation or the principle." See Bonnet's *Mémoires*, 138–9, 152–5.

Laws are based on real relations found to exist between real, individual beings; they do not have their basis in abstractions like "universality". The most reliable formulation of laws thus follows upon observations of nature rather than upon mere speculation: that is the thrust of Bonnet's definition of relations. One can see to what extent the discovery of laws falls within the purview of natural history, and why it is the pursuit of those relations "that must occupy the naturalist". "To seek an explanation of a thing is thus actually to seek the secret relation linking this thing to others. It is not simply to imagine, much less to divine. It is to compare and reconcile facts . . . ". *Tableau*, arts. V and VI, 76–7.

Bonnet gives his definition of relations its principal formulation in *Essai analytique*, § 40: "I intend in general, by *relations*, those qualities, those determinations by virtue of which different beings conspire to the same goal, or concur to produce a certain effect. This effect is a *law of nature*. Thus laws are in general the results of relations that exist between things . . . . Laws are invariable because the determinations from which they emanate are invariable. Beings are what they are: their essence is immutable." As evidence both of the central importance of relations to his work and of his conscious rejection of Montesquieu's definition, the final chapter of the *Essai analytique*, entitled "Observations on some Passages from *Esprit des Lois* Relative to this Analysis", is given over, "with discomfort at publishing these observations", to a critique of Montesquieu's definition of laws. "Are laws relations?" Bonnet asks (§ 856), immediately replying that they are not; he repeats the formulation of § 40, repeats the example of the magnet, points out that Montesquieu failed to define relations whereas he had not, and concludes, as expected, that Montesquieu should have defined laws as consequences or results of the relations existing between things.

21 Bonnet, *Essai analytique*, § 518: "The understanding does not *create* relations."

22 See e.g., *ibid.*, § § 529–30, where Bonnet tries to deny all creativity in Newton's laws of planetary motion. "If Newton has appeared to create, it is a vulgar view of creator." Instead, he has *discovered* "the relations connecting truth." Everything was deduced from observation, derived through an application of the attention. Attention replaces creativity: that is its privileged status in Bonnet's psychology. It is attention, not creativity, that is "the mother of genius". In *Analysée abrégée de l'essai analytique*, Bonnet calls for a long-overdue history of the attention, which would be, precisely, a history of all true discoveries (arts. XX, XXI, 54–60). The same point is made in *Essai analytique*, § 279, where he adds: "The works of the most celebrated observers can be regarded as memoires in the service of the history of the attention."

23 Bonnet, *Philalethe*, Chapter VI.

24 Bonnet, *Essai analytique*, § 519.

25 Bonnet, *Philalethe*, Chapter VI. My emphasis.

26 Bonnet, *Essai analytique*, § § 298–302; *Philalethe*, Chapter VI.

27 Bonnet uses the two interchangeably, at least in the *Philalethe*. See *ibid.*, Chapter I.

28 *Ibid.*, Chapter VII.

29 *Ibid.*

30 *Ibid.*, Chapter XV.

31 Bonnet, *Essai analytique*, § § 282, 305, 371, 518.

32 Bonnet, *Philalethe*, Chapter VI.

[33] Bonnet, *Essai analytique*, § 279.
[34] *Ibid.*, § 284: "A judgment is the perception of the relation that exists between two or more things."
[35] *Ibid.*, § § 288–9, 306.
[36] Bonnet, *Mémoires*, 45. Cf. *Considerations*, Chapter XII, art. 213.
[37] Bonnet, *Mémoires*, 45.
[38] Bonnet, *Essai analytique*, § 325. For the following quotations in this paragraph, see *ibid.*, § § 325–7, 332–6.
[39] Bonnet, *Philalethe*, Chapter VI.
[40] *Ibid.*, Chapter VII.
[41] *Ibid.*
[42] *Ibid.*
[43] *Ibid.*
[44] *Ibid.*, Chapter VI.
[45] *Ibid.*, Chapter X. Raymond Savioz, rather badly misunderstanding the place of analogy in Bonnet's thought, groups it under "Subsidiary Methods" and then criticizes it, in that capacity, as a method resting too heavily on reason, too little on experimentation. Savioz tries to present Bonnet as a straightforward empiricist who suffers occasionally from inexplicable lapses. *La Philosophie de Charles Bonnet de Genève*, 325–7.
[46] Bonnet, *Philalethe*, Chapter X. My emphasis.
[47] *Ibid.*, Chapter XV.
[48] *Ibid.*
[49] See Michel Foucault, *The Archaeology of Knowledge* (New York, 1972), 8–10, 12–14, 28.
[50] Bonnet, *Principes philosophiques* in *Oeuvres*, XVII, 330.
[51] Bonnet, *Contemplation*, part I, Chapter VII.
[52] It should be mentioned that sometimes the various systems are presented in a more mechanical guise, which does not really support the image of concentricity. In these cases they are likened to small gears making up a large machine: "The general system is made up of the assemblage of particular systems, which are like the different wheels of the machine. An insect or plant is a particular system, a tiny wheel that causes the greatest to move." From *Traité d'Insectologie*, quoted in *Mémoires*, 81.
[53] Bonnet, *Contemplation*, part III, Chapter V, note 2.
[54] Thus he says, in the "Préface" to the *Contemplation*, as if to warn against all misunderstanding, "In sketching the chain of beings, I have in no way pretended to fix the gradations of nature. As I have already said, it is only a way of envisaging beings and of surveying them. There are undoubtedly gradations in nature: the ancients have noted them; some of them we discover at a glance .... But the nature, order or enchainment of these gradations are only very imperfectly known to us." Cf. *ibid.*, part VIII, Chapter XVII.
[55] To be able to acquire complete knowledge of the mechanism of a simple fiber, e.g., we would then know how the entire body is nourished and grows, and from this knowledge alone we would have learned more about animal organization than has been learned through all the discoveries of modern physiology. But the elements remain enveloped in enigma, as they have despite the best efforts of mankind for the past 3,000 years. And why? Because all the parts of the organic body are so tightly enchained to one another that ignorance of the least part leads, of necessity, to obscurity for the whole system. We

suffer the same failure when examining the simplest element as we do when attempting to perceive that most subtle interplay of repetitions and differences determining the true, the infinitely nuanced, order and arrangement of terrestial beings. It is all innately beyond our grasp, beyond the scope of our vision, deeper than the penetration of our gaze. "We cast a glance upon the exterior of beings, we traverse only their outer surface." *Contemplation*, part II, Chapter XIII. For our limited knowledge of the mechanism of the simple fiber, and its implications, see *Palingénésie*, part XII, Chapters I–IV. Reflecting on the limits of our faculties, the imperfection of our methods, and the vast number of objects that remain beyond the scope of our senses, Bonnet is led to his rather novel conclusion "that this world which we inhabit has not been created principally for us", but, rather, for higher intelligences who are endowed with the faculties necessary to embrace the economy of the terrestrial system. *Ibid.*, Chapters II, VIII; *Contemplation*, part IV, Chapter XIII.

CHAPTER VI

CONCLUSION

"What is an idea?" Voltaire asked rhetorically in his *Philosophical Dictionary*. "It is an image painted in my brain . . . ideas are nothing but the consequences of all the objects I have perceived .... I have ideas only because I have images." The idea as a representative image of the thing, this was the view everywhere accepted in the eighteenth century. Representation served as the foundation of all knowledge, while "observation", so much revered in any discussion of method, clarified the representation, enhancing the precision and accuracy of the reproduction. The foundation of language, however, and the awakening of reflective thought rested on the *duplicated* representation of the arbitrary sign. Whereas the representation fulfilled its purpose precisely to the extent that it provided an accurate reproduction of the thing, this duplicated representation was valued for its very artificiality, for its detachment. Its function could hardly have been more clearly expressed than in Bonnet's formulation, "the more indeterminate the sign, the more it is a sign; for it has greater representative capacity".

Through the artificiality of its assignation, the linguistic sign holds up the representation for further comparison, judgment, modification, generalization or specification. It provides the mind a vehicle by which to traverse its representations, and by this means activates the incisive edge of analysis: representations may be carved up, the elements examined, categorized and reassembled however the individual undertaking these procedures may wish. Language makes this possible. Representations are impressed upon us randomly and automatically, but signs permit us to make use of these representations, to gain knowledge of the world and ourselves, and to gain control of our circumstances.

Yet representations would remain nothing but discrete images, even with the support of language, were it not that they arise out of a world tightly interlaced, a harmonious whole, where the connections are there to be made out, not invented. This relational order is static: all relations, even the most indirect, are preexistent, unchanging, and entirely determinate. Bonnet does of course introduce change into the world of ontological relations in the form of "world revolutions", but these events, too, are preformed and the picture remains static. Periodically they go off, like time-bombs set at the Creation, and rearrange the entire relational order. What most fascinates Bonnet is the magnitude of each reordering. As he says, any taxonomy from an earlier period would be useless. Upon the surface of the earth, a new relational network or chain is inscribed. Were we ever in a position to survey the whole history of the world, it would appear like a stack of so many taxonomic tables.

At various points Bonnet's analyses become uncertain, begin to break down and turn into something else, and then are retrieved. As in a picture that discloses one thing in the line-work and another in the spaces, Bonnet occasionally saw an image leap out from the spaces, an apparition with which he did not know how to contend. Let us, for the moment, push the arbitrary sign and the relation a step further, as Ferdinand de Saussure did in reaching back to the eighteenth century and recovering the arbitrary sign,[1] and trace this apparition. To Bonnet and his contemporaries, language was a nomenclature labeling the representations it could not but accept. The sign's "arbitrariness" extended only to the word chosen, not to the concept represented. To question the representation was the beginning of trouble. Thus it was so often said that we "create nothing" in thought, an avowal repeated even by those, like Bonnet, who believed in the active powers of the mind. Every idea had to be connected to the world; none could be allowed to spring autochthonous from the mind, for that put them all in doubt. In the discussion of relations, too, Bonnet clung to the notion that relations were grounded in essences, that things or beings had their "essential" properties from which relations derived and which they reflected. All things were joined through the relations holding between their "true natures." Beings cast their images in our senses, and we tried to uncover, in those representations, the essential properties and essential relations.

Saussure, however, redefines the arbitrary sign to include both the signifier (word) and the signified (concept). He begins by allowing what posed only a problem and a boundary to eighteenth-century writers, when they saw it at all. If the sign includes the concept, then language has become something other than the naming of stable and preexistent things. Language organizes the world in the very process of articulating it. Concepts are arbitrary divisions within a linguistic system. Saussure no longer tries to assure himself that the signified stands, in some way, as the autonomous representation of essential, defining features of beings and objects in the world, or of their relationships to each other and to us. The signifieds are simply members of a system and are defined by their relations to other members. These relations are not referred back to an exterior order; they are referred back only to the linguistic system itself.

Bonnet was not entirely ignorant of this perspective. He did not exclude, from his reflections, an uncertainty about the representative accuracy of our thoughts, of our language, or of our classifications of nature. But he did not doubt that our mode of thought is anything other than representational, or that a fundamental relational order of things stood as the basis of all our apprehensions and misapprehensions.

Historians have failed to understand that the chain of beings was not presented by Bonnet as a description of the actual order of nature; rather, it served as a model or as a metaphor, useful for taxonomic purposes as a means of envisaging a configuration of all the relations binding together everything in the world. It was these relations that were foundational for him. It was their nature and scope that had to be understood, both among the objects of the world and among the objects of the mind, in order that some semblance of the interrelated whole might be grasped. Now, homologous to the chain-of-beings metaphor, binding the world of objects together in an organizational framework (ontological relations), the metaphor of the statue-man performed the same function with regard to the world of ideas (epistemological relations). Both models made it possible to submit the interconnected whole to scrutiny – to the scissions of analysis and the recombinations of synthesis: that is, to the performance of logical operations and to the ordering of identities and differences across taxonomic tables. All of which shows the statue-man and the chain of beings as having emerged from a particular epistemological arrangement, and, with the context of contemporary philosophical concerns, to have been employed as useful devices.

Scanning the sites of our probes into Bonnet's texts, it may be observed that we have never departed from epistemological issues, that the nature of human understanding followed Bonnet through the construction of his taxonomy, through his speculations on language, education and personality, and dominated (as well as troubled) his thoughts on the relational order of the universe. But these issues, so obviously concerned with knowing (and at times with the limits of reason), have been shown to be equally linked to language, and in particular to the natural-conventional system of signs. From Port-Royal to the end of the eighteenth century, language embodied ruled thought, displayed and disseminated order through its own "natural" means. Thanks to language there could exist beings distributed in taxonomies, relations arranged into configurations and, in fact, representations of every kind unfolded into the linear sequence of sentences. Hence, despite the variety of topics we have considered under Bonnet's system of thought, it can now be seen that we have never really left the domain of language. We apply signs to the world of beings, to our world of thought, and by this means take control of ourselves – in a world in which most creatures are tossed on a sea of sensations. We thereby fulfill the level of our present perfection, the dignity of our rank: a dignity that results from our ability to impose rational order(s), and an ability that is ours only because we speak.

How are we finally to traverse and put into order all the relations of knowledge which, between the object and the idea, or again, between the most sensible ideas and the most abstract, circumscribe the dominion of representative thought? In the end, this must be done through language. The compositeness of representative images must always be separated – like the surface of a map, the parts of a clock, the organs of the body – into component units, relations of various kinds determined and judged, and then reassembled into another sort of unity: analytic and linear unity, which assembles the components into a sequential order where they are named, defined and remembered; where their various systems and relations are noted; where the unraveled whole is spread across sentences and paragraphs and pages, and rejoined piece by piece in the duplicated representation of discourse. We convert our ideas into books. Were our penetration greater we could do the same with the very movements of our sense fibers producing these images: "an intelligence capable of observing these movements in the brain could read them like a book".[2]

And how, finally, do we address ourselves to the objects that have excited the representations by which, at second and third remove, we attempt to grasp the relations joining all things? But we have only the one procedure, and it is these relations we are trying to determine when we analyze our ideas. We attempt to reconstruct the harmonious whole from an already fractured image, for we have only discrete ideas, selected somewhat randomly by limited senses, from which to reconstitute reality's magnificent unity. In the end we again have recourse only to language and the method it has given, through which we examine the various characters of beings, rank them according to criteria of relative perfections, distribute the world of being – from the lowliest and least organized form of matter to the angels – across taxonomies in a single encyclopedic effort to incorporate everything, leaving no gaps, and hence record, by means of this exhaustive denomination, all the relations constituting universal harmony. Our efforts will of necessity remain partial.

I delight in envisaging the innumerable multitude of Worlds as so many books which, when collected together, compose the immense Library of the Universe or the true

Universal Encyclopedia. I conceive that the marvelous gradation that exists between these different worlds facilitates in superior intelligences, to whom it has been given to traverse or rather to read them, the acquisition of truths of every kind, which it encompasses, and instills in their understanding that order and that concatenation which are its principal beauty, and without which there is no true science. But these celestial Encyclopedists do not all possess the Encyclopedia of the Universe to the same degree; some possess only a few branches of it, others possess a greater number, others grasp still more; but all have eternity in which to increase and perfect their learning and develop all their faculties.[3]

In this final thought we see the great advantage of the personality that does not die. Its faculties develop and its learning increases forever.

The book of our representations, the universal encyclopedia of our being: knowledge is ordered according to the dictates, the "advantages" of language. Not that language is prior to knowledge. They arise together: "It is by the same processes that one learns to speak and that one discovers either the principles of the world's system or those of the mind's operations, that is, all that is sublime in our knowledge."[4] Nor do their interests differ. The focus of both is the analysis of representation: language unfolds the unitary representations; the reflective processes of understanding compare, judge and grasp the relations bared. It is an arrangement, as we have plainly seen, that takes in Bonnet's considerations on nature – his sceintific knowledge – as well. "This", Foucault claims,

is because it is of the very nature of science to enter into the system of verbal communications, and of the very nature of language to be knowledge from its very first word. Speaking, enlightening, and knowing are, in the strict sense of the term, *of the same order*. The interest shown by the Classical age in science, the publicity accorded to its controversies, its extremely exoteric character, its opening up to the uninitiated, Fontenelle's popularization of astronomy, Voltaire reading Newton, all this is doubtless nothing more than a sociological phenomenon. It did not provoke the slightest alteration in the history of thought, or modify the development of knowledge one jot. It explains nothing, except of course on the doxographic level where it should be situated; but its condition of possibility is nevertheless there, in that reciprocal kinship between knowledge and language.[5]

The system of Bonnet's thought thus comes together squarely within the eighteenth-century framework of problems, possibilities and solutions. If one is to understand what was meant by reason, in the "age of reason", the proper formulation appears to lie here: "Speaking, enlightening, knowing are ... *of the same order*." In which case the foregoing study of Bonnet's thought has been only a long reflection on that question; and if, to trace the boundaries and locus of reason, it had to be followed through the distribution of organized beings, the structure of personality, the operations of reflective thought, and into the heritage and control of thought dispensed in education and made possible through the gift of language, we looked for it only in the same places it was sought and delivered up in that period. What reason had to do with, specifically, was *order*. It had to do with the establishment of an order of ruled thought and with the determination of an order of the relations of being. One hoped, in the eighteenth century, to ascertain the actual order of things, the natural order. But one expected, at least to be able to establish *an* order. For as rational, speaking beings, ordering was the prerogative and the duty of our station.

NOTES

[1] Ferdinand de Saussure, *Course in General Linguistics*, trans. Wade Baskin (New York, 1966).

[2] Bonnet, *Essai de psychologie*, "Introduction", 2.

[3] *Contemplation*, part IV, Chapter XIII, note 1.

[4] Antoine Louis Claude Destutt de Tracy, *Éléments d'idéologie*, reprint (Paris, 1970), I, 24.

[5] Foucault, *The Order of Things*, 88–89.

BIBLIOGRAPHY

A. CONCERNING BONNET'S COLLECTED WORKS

Bonnet prepared the general edition of his *Oeuvres* himself, and claimed it to be the most extensive and tiring of all his literary toils. This came about when Samuel Fauché, representing Neuchatel, approached him in the Autumn of 1775 with the proposal that his works be collected. Bonnet had little enthusiasm for the project, since he neither wished to release such an edition without making substantial revisions and additions, nor looked forward to the work involved. Fauché tried to encourage Bonnet with the argument that simple lists of *errata* would suffice, but eight years later the new additions had totaled some 1,600 pages. The two works most extensively revised were the *Contemplation de la nature* and the *Essay analytique sur les facultés de l'âme*; the first took two years to revise, and the second received corrections and revisions nearly as extensive. Little was changed in the *Palingénésie philosophique*. Bonnet included the *Essai de psychologie* in his collected works, along with a foreword openly acknowledging its authorship for the first time, thus finally "providing this little bastard with a father." Also included were a number of shorter essays of varying length, none of which had previously been published: among them the *Philalethe ou essai d'une méthode pour établir quelques vérités de philosophie rationnelle*, *Recueil de divers passages de Leibnitz sur la survivance de l'animal*, *Vue du leibnitianisme*, *Nouvelles considérations sur les bornes naturelles de nos connaissances, Lettre au sujet du discours de M. J. J. Rousseau sur l'origine et les fondements de l'inégalité parmi les hommes, Remarques sur le sentiment de Clarke, touchant la liberté, Idées sur l'art d'étudier et sur l'ordre et le but des études de philosophie rationnelle, Hypothèse sur l'âme des bêtes et leur industrie, Idées sur l'origine du mal*, and *Méditation sur l'origine des sensations et sur l'union de l'âme et du corps*.

Two editions of this collection of Bonnet's *Oeuvres* were made at the same time, a larger one in quarto (10 volumes) and a smaller in octavio (18 volumes); they were published in three installments between 1779 and 1783. The *Mémoires autobiographiques*, from which this information has been taken (see 353–364), were written by Bonnet as *Mémoires sur ma vie et sur mes écrits, en forme de lettres à MM. de Haller, Trembley et de Saussure* between 1775 and 1791. They were addressed to these three people successively because first Haller and then Trembley died before Bonnet had finished. The memoirs lay in manuscript in the Geneva library until Raymond Savioz published them in 1948. For a complete list of Bonnet's published works, as well as a list of his principal unpublished letters and manuscripts, see the bibliography at the end of Savioz's *La Philosophie de Charles Bonnet de Genève*.

B. SELECTED BIBLIOGRAPHY

Aarsleff, Hans, 'The Eighteenth Century, Including Leibniz', *Current Trends in Linguistics*, Thomas A. Sebeok (ed.), *et al.*, The Hague: Mouton (1974).

Aarsleff, Hans, 'The History of Linguistics and Professor Chomsky', *Language* **46**: 3 (1970), 570–85.

Aarsleff, Hans, 'Leibniz on Locke on Language', *American Philosophical Quarterly* **I**: 3 (July 1964), 165–88.

Aarsleff, Hans, *The Study of Language in England, 1780–1860*, Princeton: Princeton University Press (1967).

Aarsleff, Hans, 'The Tradition of Condillac', *Studies in the History of Linguistics*, Dell Hymes (ed.), Bloomington: Indiana University Press (1974).

Archer, R. L. (ed.), *Jean Jacques Rousseau: His Educational Theories Selected from Émile, Julie and Other Writings*, Woodbury, New York: Barron's Educational Series (1964).

Arnauld, Antoine, *The Art of Thinking: Port-Royal Logic*, Trans. James Dickoff and Patricia James. New York: Bobbs-Merrill Company (1964).

Auroux, Sylvain (ed.), *L'Encyclopédie 'Grammaire' et 'Langue' au XVIIIe siècle*, Paris: Maison Même (1973).

Axtell, James L. (ed.), *The Educational Writings of John Locke*. Great Britain: Cambridge University Press (1968).

Bachelard, Gaston, *The Psychoanalysis of Fire*, Trans. Alan C. M. Ross. Boston: Beacon (1964).

Barthes, Roland, *Elements of Semiology*, London: Jonathan Cape (1967).

Barthes, Roland, *S/Z*. Trans. Richard Miller. New York: Hill and Wang (1974).

Bowler, Peter J., 'Bonnet and Buffon: Theories of Generation and the Problem of Species', *Journal of the History of Biology* **6**: 2 (Fall 1973), 259–81.

Bowler, Peter J., 'Evolutionism in the Enlightenment', *History of Science* **XII** (1974), 159–83.

Boyd, William, *From Locke to Montessori*. London: G. G. Harrap and Co. (1917).

Brosses, Charles de, *Traité de la méchaniques des langues et des principes physiques de l'étymologie*, Paris: Terrelonge (1801).

Brunot, Ferdinand, *Histoire de la langue française des origines à 1900*, VI (*Le XVIIIe siècle*), Paris: Armand Colin (1932).

Canguilhem, Georges, *La Connaissance de la vie*, Deuxième éd., revue et augmentée. Paris: J. Vrin (1969).

Cassirer, Ernst, *The Philosophy of the Enlightenment*, Trans. Fritz C. A. Koeller and James P. Pettegrove. Boston: Beacon (1955).

Caton, Hiram, *The Origin of Subjectivity*, New Haven: Yale University (1973).

Claydon, Leslie F. (ed.), *Rousseau*, London: Collier-Macmillan Limited (1969).

Condillac, Etienne Bonnot de, *Essai sur l'origine des connaissances humaines. Oeuvres complètes*, New ed. Paris: Baudouin Frères (1827), I.

Coleman, William, *Biology in the Nineteenth Century: Problems of Form, Function, and Transformation*, New York: John Wiley and Sons (1971).

Cranston, Maurice, *John Locke: A Biography*, London: Longmans, Green and Company (1957).

Crocker, Lester, *An Age of Crisis: Man and World in Eighteenth Century French Thought*, Baltimore: Johns Hopkins (1959).

Culler, Jonathan, *Ferdinand de Saussure*, New York: Penguin (1977).

Cuvier, Georges, *Histoire des sciences naturelles* IV. Paris: Fortin (1843).

Cuvier, Georges, *Recueil des éloges historique*, I, Paris: F. G. Levrault (1819).

Deleuze, Gilles, *Différence et Répétition*, Paris: Presses Universitaires de France (1972).

Derrida, Jacques, 'Differance', *Speech and Phenomena and other Essays on Husserl's Theory of Signs*. Trans. David B. Allison. Evanston: Northwestern (1973).

Derrida, Jacques, *Of Grammatology*, Gayatri Chakravorty Spivak (transl.), Baltimore: Johns Hopkins (1976).

Derrida, Jacques, 'Structure, Sign and Play in the Discourse of the Human Scineces', *The Languages of Criticism and the Sciences of Man*, Richard Macksey and Eugenio Donato (eds.), Baltimore: Johns Hopkins (1970).

Destutt de Tracy, Antoine Louis Claude, *Éléments d'idéologie*, Paris: J. Vrin (1970).

Diderot, Denis, *Rameau's Nephew and Other Works*, Trans. Jacques Barzun and Ralph H. Bowen. New York: Bobbs-Merril (1956).

Dijksterhuis, E. J., *The Mechanization of the World Picture*, Trans. C. Dikshoorn. New York: Oxford University Press (1961).

Eiseley, Loren, *Darwin's Century: Evolution and the Men Who Discovered It*, Garden City, N. Y.: Doubleday (1958).

Farber, Paul L., 'Buffon and the Concept of Species', *Journal of the History of Biology* **5**: 2 (Fall 1972).

Formigari, Lia, 'Language and Society in the Late Eighteenth Century', *Journal of the History of Ideas* **XXXV**: 2, (April–June 1974), 275–92.

Foucault, Michel, *The Archaeology of Knowledge*. Trans. A. M. Sheridan Smith. New York: Pantheon, 1972.

Foucault, Michel, 'La grammaire générale de Port-Royal', *Languages* 7 (September 1967), 7–14.

Foucault, Michel, *The Order of Things: An Archaeology of the Human Sciences*. New York: Pantheon (1970).

Gasking, Elizabeth, *Investigations into Generation, 1651–1828*, Baltimore: Johns Hopkins, n.d.

Gay, Peter (ed.), *John Locke on Education*, Classics in Education, No. 20. New York: Columbia University Press, (1964).

Gay, Peter, *Voltaire's Politics: The Poet as Realist*, New York: Vintage (1965).

Gillespie, Charles Coulston, *The Edge of Objectivity: An Essay in the History of Scientific Ideas*, Princeton: Princeton University Press (1960).

Gillespie, Charles Coulston, *Genesis and Geology*, New York: Harper Torchbooks (1959).

Glass, Bentley *et al.*, (eds.), *Forerunners of Darwin: 1745–1859*, Baltimore: Johns Hopkins (1959).

Gotshalk, D. W., *Structure and Reality: A Study of First Principles*, New York: Dial Press (1937).

Gould, Stephen Jay, *Ontogeny and Phylogeny*. Harvard: Belknap Press of the Harvard University Press, (1977).

Harrison, John and Laslett, Peter. *The Library of John Locke*, The Oxford Bibliographical Society. NS XIII. Oxford: Oxford University Press (1965).

Hintikka, Jaakko, 'Leibniz on Plenitude, Relations and the 'Reign of Law',' *Leibniz: A Collection of Critical Essays*, Harry G. Frankfurt (ed.), Garden City, N. Y.: Doubleday (1972).

Hyman, Stanley Edgar, *The Tangled Bank: Darwin, Marx, Frazer and Freud As Imaginative Writers*, New York: Grosset and Dunlop (1966).

Ishiguro, Hidé, *Leibniz's Philosophy of Logic and Language*, New York: Cornell University Press (1972).

Jacob, François, *The Logic of Life: A History of Heredity*, New York: Vintage Books (1976).

Jeffreys, M. V. C., *John Locke: Prophet of Common Sense*, London: Methuen (1967).

Josephs, Herbert, *Diderot's Dialogue of Gesture and Language: Le Neveu de Rameau*, Ohio: Ohio State University Press (1969).

Juliard, Pierre, *Philosophies of Language in Eighteenth-Century France*, The Hague: Mouton (1970).

Knight, Isabel F., *The Geometric Spirit*, New Haven: Yale University Press (1968).

Kuhn, Thomas S., *The Structure of Scientific Revolutions*, Chicago: University of Chicago Press (1962).

Larson, James L., 'Linnaeus and the Natural Method', *Isis* **58**: 193 (Fall 1967).

Leibniz, Gottfried Wilhelm, *The Monadology* in *Leibniz Selections*, Philip P. Wiener (ed.), New York: Charles Scribner's Sons, 1951.

Leibniz, Gottfried Wilhelm, *Theodicy* (ed.), Abridged and with an intro. by Diogenes Allen. New York: Bobbs-Merril (1966).

Lévi-Strauss, Claude, *The Savage Mind*, Chicago: University of Chicago Press (1966).

Locke, John, *An Essay Concerning Human Understanding*, 2 Vols. Collated and annotated by Alexander Campbell Fraser. New York: Dover (1959).

Logique de Port-Royal, Critical ed. by Pierre Claire and Francois Girbal. Paris: Presses Universitaires de France (1965).

Lovejoy, Arthur O., *The Great Chain of Being: A Study of the History of an Idea*, New York: Harper (1960).

Lyon, John, 'The 'Initial Discourse' to Buffon's *Histoire naturelle*: The First Complete English Translation', *Journal of the History of Biology* **9**: 1 (Spring 1976).

Martin, Gottfried, *Leibniz, Logic and Metaphysics*, K. J. Northcott and P. G. Lucas (transl.), Manchester: Manchester University Press (1964).

Marx, Jacques. 'Alchemie et palingégésie', *Isis*, **62**, part 3, No. 213 (September 1971), 275–89.

Masters, Roger D., *The Political Philosophy of Rousseau*, Princeton: Princeton University Press (1968).

Mettrie, Julien Offray de la, *L'homme machine*, Princeton: Princeton University Press (1960).

Mornet, Daniel, *Les sciences de la nature en France, au XVIIIe siècle. Un chapître de l'histoire des idées*, Paris: A. Colin (1911).

Nisbet, Robert A., *Social Change and History*, New York: Oxford University Press (1969).

Offner, Max, *Die Psychologie Charles Bonnets: Eine Studie zur Geschichte Der Psychologie*, Schriften der Gesellschaft für psychologische Forschung. Leipzig: Verlag von Ambr. Abel (1893).

Ollman, Bertell, *Alienation: Marx's Conception of Man in Capitalist Society*, New York: Cambridge University Press (1971).

Palmer, Robert R., *Catholics and Unbelievers in Eighteenth Century France*, Princeton: Princeton University Press (1939).

Perkins, Jean A., *The Concept of the Self in the French Enlightenment*, Genève: Droz (1969).

Popkin, Richard, 'Skepticism in the Enlightenment', *Studies on Voltaire and the 18th Century*, XXVI. Genève: Institute et musée Voltaire (1963), 1321–45.

Reardon, Michael, 'Pierre Ballanche as a French Traditionalist', *Catholic Historical Review* 53 (January 1968), 573–99.

Roe, Shirley A., 'The Development of Albrecht von Haller's Views on Embryology', *Journal of the History of Biology* 8: 2 (Fall 1975).

Roe, Shirley, A., 'Rationalism and Embryology: Caspar Friedrich Wolff's Theory of Epigenesis', *Journal of the History of Biology* **12**: 1 (Spring 1979).

Rogers, Jacques, *Les sciences de la vie dans la pensée française du XVIIIe siècle*, Paris: Armand Colin (1963); 2nd ed. (1972).

Rousseau, Jean-Jacques, *Discours sur l'origine et les foudements de l'inégalité parmi les hommes*, Paris: Gallimard (1965).

Rousseau, Jean-Jacques, *Émil: ou, de l'éducation*, Paris: F. Didot (1851).

Rousseau, *Essay on the Origin of Languages*, Trans. John H. Moran. New York: Frederick Ungar Publishing Company (1966).

Rousseau, Jean-Jacques, *The First and Second Discourses*, Roger D. Masters (ed.), Roger D. and Judith R. Masters (transl.), New York: St. Martin's Press (1964).

Saussure, Ferdinand de, *Course in General Linguistics*, Charles Bally and Albert Sechehaye (eds.) in collaboration with Albert Riedlinger. Trans. with intro. and notes, Wade Baskin. New York: McGraw-Hill (1966).

Savioz, Raymond, *La philosophie de Charles Bonnet de Genève*, Paris: J. Vrin (1948).

Savioz, Raymond (ed.), *Mémoires autobiographiques de Charles Bonnet de Genève*, Paris: J. Vrin (1948).

Schiller, Joseph, 'Queries, Answers and Unsolved Problems in Eighteenth Century Biology', *History of Science* **XII** (1974), 184–99.

Sloan, Philip R., 'The Buffon-Linnaeus Controversy', *Isis* **67**: 238 (September 1976).

Smith, D. W. *Helvetius: A Study in Persecution*, Oxford: Clarendon Press (1965).

Thomas, Lewis, *The Lives of a Cell*, New York: Viking (1971).

Vartanian, Abram, *Diderot and Descartes: A Study of Scientific Naturalism in the Enlightenment*, Princeton: Princeton University Press (1953).

Vartanian, Abram, 'Trembley's Polyp, La Mettrie, and 18th Century Materialism', *Journal of the History of Ideas* **XI**: 3, 259–86.

Whiston, William, *A New Theory of the Earth, from Its Origin to the Consummation of all Things*, R. Roberts, 1691. Xerox University Microfilms, Reel 301. Ann Arbor, Michigan, 1973.

Whitman, C. O., 'Evolution and Epigenesis,' 'Bonnet's Theory of Evolution: A System of Negation', 'The Pallingenesis and the Germ Doctrine of Bonnet', *Biological Lectures, Woods Hole*, Boston (1894–6).

Wilson, Edmund B., *The Cell in Development and Inheritance*, New York: Macmillan (1898).

Windelband, Wilhelm, *A History of Philosophy, Volume II: Renaissance, Enlightenment, and Modern*, New York: Harper and Brothers (1958).

Wohl, Robert, 'Buffon and His Project for a New Science', *Isis* **51**: 2 (June 1960).
Yolton, John, *John Locke and Education*, New York: Random House (1971).
Yolton, John, *Locke and the Compass of Human Understanding*, London: Cambridge University Press (1970).
Yolton, John, *The Locke Reader*, London: Cambridge University Press (1977).

INDEX